D0457965

ESRI Press

REDLANDS, CALIFORNIA

Thinking About GIS

Geographic Information System Planning
for Managers

Thinking About GIS

About

GIS

**Geographic Information System Planning
for Managers**

Roger Tomlinson

ESRI PRESS
REDLANDS, CALIFORNIA

ESRI
    Thinking about GIS: geographic information system planning for managers
    ISBN 1-58948-070-8
First printing
Printed in the United States of America.

Library of Congress Cataloging-in-Publication Data
Tomlinson, Roger F.
Thinking about GIS : geographic information system planning for managers / Roger Tomlinson.
    p.   cm.
    Includes bibliographical references and index.
    ISBN 1-58948-070-8 (hardcover : alk. paper)
    1. Geographic information systems. I. Title.
    G70.212.T66 2003
    910'.285—dc21   2003012141

Published by ESRI, 380 New York Street, Redlands, California 92373-8100.
Books from ESRI Press are available to resellers worldwide through Independent Publishers Group (IPG). For information on volume discounts, or to place an order, call IPG at 1-800-888-4741 in the United States, or at 312-337-0747 outside the United States.

Dedicated to Mr. Leonard Hassell, geography teacher at Newmarket Grammar School, Suffolk, England, who turned my heart and life toward geography.

R. F. T.

Acknowledgments

The errors are mine. The methods described in this book have evolved over the years with help from many people. These include the associates of Tomlinson Associates Ltd. in Canada, the United States, and Australia. Many contributions were made by my colleague Larry Sugarbaker as we developed the "Managing a GIS" seminar together. Generous input has been received from the staff of our clients worldwide, and the staff of the corporations that eventually served our clients. They have been the real-world laboratory in which the ideas were tested. This book exists because Jack Dangermond thought it was a good idea to show the world what we put him through and because Brian Parr and Christian Harder turned the methodology into readable words. Their support has been patient and constant. None of this would have come about without the continuing work and friendship of my wife Lila. Not only has she typed every word and corrected most of my mistakes, but she still smiles. To these I owe my heartfelt thanks.

Roger Tomlinson

For a number of years, Roger Tomlinson has been advocating that one of the key ingredients to successful GIS is the use of a consistent planning methodology. He has developed and evolved a methodology and adapted it over the years through his personal consultation practicum and in association with the evolution of technology. At the ESRI® International User Conference and other venues, Tomlinson has taught his method as part of a very popular "Managing a GIS" seminar. In observing the attendance of these seminars, I have noticed that they tend to attract two primary groups of people: the first group is comprised of senior managers who oversee GIS and other information technologies in their organizations. The second group is comprised of more technical managers responsible for the actual implementation of GIS and other information technologies. That these two markedly differing groups would come together year after year to glean Tomlinson's wisdom always struck me as significant, and as an excellent starting point for a book on GIS planning.

So this book is in fact written for those two kinds of managers, and is intended to bridge the communication gap between them. Senior executives in public and private sector organizations often have a general idea that GIS would be good for their organization and they know how to get the resources allocated to make it happen. What they lack is enough understanding about the capabilities and unique constraints of geospatial data technologies needed to direct and ask the right questions of their technical managers (the second audience). Conversely, these line GIS managers tend to have a solid grasp of the technology and the unique characteristics of GIS, but know much less about how the GIS must operate within the context of their larger organization. What they need is information that will allow them to anticipate the questions that their bosses are going to ask. This book effectively and successfully serves the needs of both groups.

While these are the primary audiences, the book also has value for the student of GIS who wishes to learn how to do the middle manager's job. It is an invaluable source of tuition for students to understand what being a GIS manager in a large organization is all about.

While Roger has rightly become known as the "father of GIS" as a result of his early work in using computers to model land inventories for the Canadian government in the early 1960s, I believe that his greatest contribution to the field is the rigorous method of GIS planning that is described in this book. I hope that you find his work as informative and beneficial as have my colleagues and I at ESRI.

Jack Dangermond
President, ESRI

Contents

Introduction

If you're holding this book, perhaps it's because you've been charged with launching or implementing a geographic information system, a GIS, for your organization. Yours could be the type of organization that has historically used GIS—a local government, a transportation authority, a forest management agency. Or it could be the type of organization—such as a corporation, a political action group, or a farm—that has only recently begun to discover the positive implications of geospatial computer programs.

The GIS you've been tasked with implementing could be intended to serve a single, specific purpose, or to perform an ongoing function. It could even be what's called an "enterprise GIS," one designed to serve a wide range of purposes across many departments within your organization. (And as you'll learn as your GIS evolves, a well-planned implementation can start out as a project and grow, or scale, into a full-blown enterprise system.)

Whatever the mission of your organization, or the intended scope of the initial GIS implementation, the good news is that the fundamental principles behind planning for a successful GIS are essentially the same. These principles are based on the simple concept that you must think about your real purposes and decide what output, what information, you want from your GIS. All the rest comes out of that.

This book details a practical method for planning a GIS that has been proven successful time and again during many years of use in real public- and private-sector organizations. It is a scaleable approach, meaning that it can be adapted to any size GIS, from a modest project to an enterprise-wide system.

The current GIS planning environment

It is useful to consider the impact that rapid advancements in GIS technology can have on the GIS planning process.

Improvements in usability and advanced off-the-shelf GIS functionality have created an environment wherein GIS development can take place more rapidly and also much more iteratively. What we see now is that geographic information systems tend to evolve over time. Planning is no longer a one-time event. Many of the standard and commonly used data sets are now readily available in digital form, and at much lower cost than even just a few years ago. This relative abundance of economical and reliable spatial data significantly widens the scope of potential GIS applications. CPU seconds are approaching zero cost. The location of human and data resources in the organization and communication between them are now the driving factors in system design. Rapid prototyping and development tools such as ESRI ArcGIS® with ModelBuilder™, Microsoft® VisualBasic®, and CASE technology allow for quick exploration and testing of applications. A greater range of options can be explored. Targeted planning on selected business areas is viable. Databases can be built incrementally and scaled-up as needed.

These days, most GIS handle spatial data under one of three paradigms. The first is the traditional stand-alone desktop information system that performs an integrated set of GIS functions on a wide variety of data types. The second is the developer environment in which a set of application-neutral, individual function components can be combined by software developers to create new applications. The third is the server environment. Here a set of standardized GIS Web services (e.g., mapping, data access, geocoding) support enterprise-wide applications. These environments currently interoperate, but are moving toward more unified models and interfaces in the future.

The value of GIS is seen to be increasingly important to organizations and public policy. Also clearly visible is the increasing need for coordination, collaboration, and an enterprise view of GIS management. GIS planning itself is increasingly important if these objectives are to be achieved.

Why plan?

So why should you plan? What's wrong with just buying some computers and GIS software, loading some data, and just sort of "letting things happen"? Can't you just adapt as things move along, tweak the system, learn as you go? In fact, doesn't all this advance thinking slow things down and just create even more work?

On the contrary, the evidence shows that good GIS planning leads to GIS success, and absence of planning leads to failure. Whether you are working with an existing system or creating a GIS from scratch, you must integrate sufficient planning into the development of your GIS; if you don't, chances are you'll end up with a system that doesn't meet your expectations.

The key undertaking of the manager is to understand their business and identify the GIS information products that would benefit that business. This understanding leads to identification of the data needed and the issues of tolerance and error and concepts of database design on which efficiency will depend. From the data requirements one then specifies system scope, software functionality needs, and hardware and network requirements. From these itemized necessities accurate cost models can be developed that will allow clear and meaningful benefit-cost analysis. Having laid this groundwork, issues affecting implementation—institutional, legal, budgetary, staffing, risk, timing—can be resolved. The end result is an effective, efficient, and demonstrably beneficial GIS within the organization.

Good planning will make your ongoing GIS efforts cost-effective in the long run. Implementation and maintenance can be expensive. Your GIS can quickly become a money pit if it's not creating useful products for the organization, ultimately jeopardizing the very existence of the GIS initiative, and perhaps your own job. Conversely, a GIS can prove its worth and justify its existence if it manages to help the cause by streamlining existing workflow and creating useful information products. These are the ultimate benefits reaped by any successful information system. This book will teach you to evaluate the benefits of the system relative to its cost, and how to make the case to management in a way that makes them advocates for your own success.

Knowing what you want to get out of your GIS is absolutely crucial to your ultimate success. Too often, organizations decide they want a GIS because they've heard great things from their peers in other organizations, or they just don't want to get left behind technologically. So they invest considerable sums of money into technology, data, and personnel without knowing exactly what they need from the system. That's like packing for a vacation without knowing where you're going. You pack everything from your closet, just in case, but it turns out that sweater isn't needed in Fiji, and you forgot the sunscreen. You've wasted time and energy and, worse yet, you're still not ready. When you try to develop a GIS without first seriously considering the real purpose, you could find yourself with the wrong (expensive) technology and unmet needs.

You must determine your organization's GIS needs from the outset of the planning process. GIS has many potential applications, so it's important to establish your specific requirements and objectives from the beginning. That way, you will avoid the chaos that results from trying to create a system with no priorities or ends in mind.

The methodology described in these pages will show you how to describe and prioritize the GIS needs of your organization in order to plan a system that meets these requirements. The entire planning process can take some time, and you may find that some of the steps can be minimized or eliminated in certain situations. But it is nonetheless important to think carefully about each step to really "get your head around" the subject. You'll be glad you took the time.

GIS: the whole picture

"No GIS can be a success without the right people involved. A real-world GIS is actually a complex system of interrelated parts, and at the center of this system is a smart person who understands the whole."

GIS: the whole picture

GIS is a particulary horizontal technology in the sense that it has wide-ranging applications across the industrial and intellectual landscape. For this reason, it tends to resist simplistic definition. Yet the first thing we need is a common understanding of what we're talking about when we refer to GIS. A simple definition is not sufficient. In order to discuss GIS separate from the context of any specific industry or application, we need a more flexible tool for elaborating: a model.

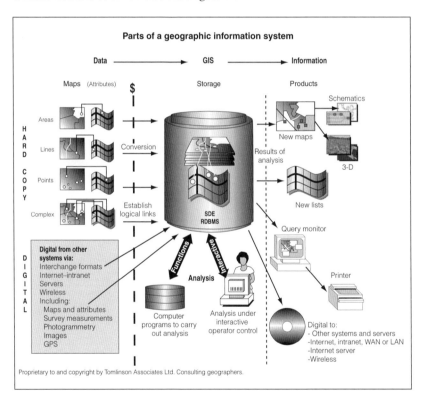

The illustration presents a holistic model of a functional geographic information system. It shows that GIS stores *spatial data* with logically-linked attribute information in a *GIS storage* database where analytical functions are controlled interactively by a human operator to generate the needed *information products*.

Let's understand the model better by examining its individual components. *Spatial data* is a term with special meaning in GIS. It is the raw data that makes the rest possible. Spatial data is distinguished by the presence of a geographic link. In other words, something about that piece of data is connected to a known place on the earth, a true geographic reference. Features commonly found on a map—roads, lakes, buildings—are also commonly found in a GIS database as individual thematic layers. Most can be represented using a combination of *points*, *lines*, or *polygons*. A crucial and related form of spatial data is data about those roads, lakes and buildings. These are called *attributes* in GIS parlance, and, in fact, it is the richness and depth of these attributes that make spatial data so potent in the context of a GIS. So where does this data come from?

Not surprisingly, good old-fashioned paper maps and other hard copy records still supply much of the physical and human geographic information needed for GIS. After all, printed paper maps have been the standard vehicle for conveying geographic information since the earliest recorded history. By scanning and digitizing the features drawn on our organization's paper maps, we mine this rich data source. And by establishing logical links to other digitized hard copy records in our organization—tables, lists, documents—we further convert data for use in the GIS, doing this until we've digitized and linked all the relevant paper documents at our disposal. More and more spatial data is also now available already in digital form. Sometimes it is available via data-sharing arrangements, sometimes it can be purchased.

Measuring and survey devices, including GPS receivers, photogrammetry images, and survey instruments generate troves of GIS-usable data. The Internet and the availability of common interchange formats also provide a means for moving all this data around quickly. All this data is systematically integrated with logical links into the database in your GIS under the primary organizing key of geographical location.

All this linked spatial data resides in the *GIS storage* system where it is available for functions such as analysis and mapmaking. The power of the computer is used to ask questions of the spatial data, to search through it, compare it, analyze it, and measure it. You use the GIS to do things that it would be very laborious or even impossible to do in any other way. These functions are under the interactive control of the GIS operator whose job it is to create the needed information products.

The information products include new spatial and attribute data, new lists and tables, on-screen queries, hard copy maps and reports, and digital information. This is the harvest, the accomplishment which represents the ultimate success of your GIS.

Identifying those end-product reports is central to the GIS planning process.

Scope of GIS projects

The three levels of scope that can be defined to categorize any GIS implementation are: single-purpose projects, department-level applications, and multidepartment enterprise systems. The same guiding principles of GIS planning apply to all three. There are subtle differences between the project types, and ultimately some of the planning steps may not be needed on small projects or department-wide applications. Understanding the scope of your own project will help you develop an effective plan for your own GIS implementation.

Most organizations end up testing the GIS waters with a *single-purpose project* carried out within a single department of an organization. The expected result is a project-specific output, such as information needed to make a decision. These can be one-time efforts that have an end date. The acquisition cost is paid for by the project, and no long-term support is expected. A site analysis to locate a new landfill would be an example.

A *department-wide application* supports at least one important ongoing business function inside an organization. An established need exists. This makes it easy to identify the set of information products that can be output related to that business objective or function. The department responsible for that business activity manages the system. Ongoing financial support is critical and corporate support is important, but not necessarily essential. The objective is not to produce long-term lasting strategic direction or support for the organization, but to support a single established business need. An example would be a city planning department that uses GIS to generate mailing lists of property owners within 300 feet of a specified property any time a change in land-use zoning is proposed for that property. The GIS would be located right there in the planning department, not elsewhere in the organization (for example within the IT division). Ongoing support is required for the application to cover staff, hardware, software, applications, and maintenance. Corporate support is desirable as they will approve the funding, but the support of the departmental head is most important because of the reliance on departmental resources to develop the application.

Finally, *enterprise-wide systems* are those that allow the organization's members to access and integrate GIS data across all departments. Enterprise-wide GIS supports departmental business needs and strategic business decisions for multiple departments. The GIS becomes a powerful tool inside the organization: the GIS is in alignment with the strategic direction of the organization and supports strategic business decisions that have been recognized within the organization. Along with long-term support from multiple departments, corporate management support is essential to enterprise-wide GIS systems. An example would be the transportation company that uses multiple GIS applications and huge databases of geospatial data across all departments as a mission-critical element of the operating strategy of the business.

The data required to support the work will come from multiple divisions and require corporate support. An enterprise-wide GIS allows the integration of this data with the business functions and processes.

The power of GIS is most leveraged at the enterprise-wide level. It is here that consistent information is available across the organization. Organizational decision makers get a clear picture of reality, data is consistently updated, more is shared, a wider-cast information net allows for better decision making, and duplication of effort is avoided.

Industry observers will tell you that the next step is societal wide—where every functioning part of society will find that they make use of GIS in the same way that they make use of computers today. Thanks to the Internet, GIS is currently evolving from the enterprise to society.

The who, what, when, where, why

Let's borrow a page from the reporter's notebook and set this story up via those famous "W"s of the newsroom: Who, What, When, Where, and Why?

Who should plan a GIS? The GIS manager must take the lead role in the planning process, but he should never go it alone. He must keep the senior-level decision makers advised and informed throughout the process. Failure to keep these budget keepers apprised can lead to reduced or eliminated funding. The only way to keep the decision makers on board is to keep them educated and actively engaged in the planning process. In fact, you will ultimately have to rely on them to tell you what information products are going to be needed at their level. The other crucial people to include in the planning process are those who will be using the system. If they aren't involved, you'll probably fail to meet their real needs.

A note on consultants: if you decide to hire GIS consultants, have them lead you through the planning steps—never hire a consultant to do the planning for you. You and your colleagues need to do the planning—it's your GIS system, your job, and your reputation at stake. As a team, you'll have to make decisions through the planning process. If you do use a consultant to help you, there should still be a GIS team in the organization that will carry out the work under the consultant's guidance. In Canada there is a saying:

"Consultants disappear like the snow in springtime."

The point is that at the end of the process, you will be left with the system to implement. If you haven't been totally involved in planning and writing the implementation strategy, this could become a very difficult and painful time.

What to plan? Since GIS is a complex system of interconnected parts, it should not be any surprise that there are six different major components that must be considered in any GIS plan:

Information products: These are the desired output from the GIS. Output may take the form of maps, reports, graphs, lists, or any combination thereof. Identifying these products with sufficient clarity early in the planning process is critical to your mission.

Software: These are the computer programs that provide the functions needed to perform analysis and create the information products desired. Sometimes there is customized software that sits on top of the main GIS software package. Updates need to be planned to keep the versions current. There are also support and operating system issues related to software.

Data: By knowing what information products you want, you can plan for the acquisition of the needed data. What can you get that already exists? What can you create from existing sources? What levels of map accuracy and scale will you need? And don't forget data format, a question that is related to the second component above, software. Sometimes data format alone can drive the software decision, as in the case of the city municipal government that wants to enter a data-sharing arrangement with the GIS department of their county who uses a certain software.

Hardware: GIS is demanding of hardware. You must take a hard look at your organization's computational resources and upgrade accordingly to support GIS. Typically a few powerful workstations support the heavy lifting and geoprocessing while "thin" clients on the network provide the user access for query and display of the GIS database. A robust internal network and fat pipe to the Internet are also required to facilitate file sharing, data acquisition, and reporting.

Procedures: Procedures are an important component and should not be left off the planning to-do list. This concerns the way that people do their jobs, and the changes that people will need to make in order to do their jobs using your new GIS system. You need a plan for transitioning from the old way to the new way, plus you need to address how the existing legacy systems will co-exist (or not) with the GIS.

People: GIS is a thinking process that requires the right people. Will you need to hire people or do the right ones already exist? How will you hire and train and keep the staff needed with the specialized skills it takes to use or build your system? Over time, this will be your single biggest cost component.

When to plan? You plan at the beginning. But the planning process also continues after the GIS is installed. Successful GIS projects attract positive attention, which means that before long people will identify other things that they'd like to see from the GIS. This will cause you to revisit the planning process. Armed now with the empirical experience of a functioning GIS, each new iteration of the planning process becomes better calibrated to the real world. There is also a certain order of planning. An obvious example is the fact that you cannot plan for which data to acquire until you've identified which information products will be required.

Where to plan? GIS planning, to be most effective, must be carried out in the business world, not from the vacuum of your office. Because GIS has the potential to create common connections between disparate things, it is inherently a horizontal technology that can touch literally every person in an organization if the company's leaders want it to (and many do). The thorough and diligent GIS planner must meet people in the organization to learn what information they need and how they need to get it. Only by taking the time to witness people doing their work can the GIS planner ever aspire to true understanding of the business processes. And how can a planner create anything of use to anyone without this knowledge?

Why plan GIS? Good planning leads to success and poor planning leads to failure. This is true whether you are starting from scratch or building from an existing GIS. It seems obvious on the surface. But time after time GIS projects do fail, and when they do it can almost always be traced back to poor planning. Like any complex information system, GIS implementation and maintenance are costly. Every component of the system—the data, the software, the hardware, the staff—all cost an organization dearly. Attention to cost-effectiveness is needed. Rightly, no organization should support a system if it's not producing useful information products which have a positive benefit-cost.

Chapter two

Overview of the method

"Like a good roadmap, an overview of the method lets you know where you are going."

Overview of the method

The planning methodology introduced in this book will show you the steps needed to prepare for GIS planning, to assess what your requirements are, which system will meet your needs, and how to implement the system in your organization once you get the approval.

What has evolved from years of experience in planning large and small implementations in public- and private-sector companies is a 10-stage GIS planning methodology. The size and nature of your organization will determine which of the component stages are most relevant to your situation. A full enterprise-wide implementation almost certainly requires all the stages to be undertaken in full while in other cases, it may be possible to complete some stages quickly or even eliminate some stages. It is important, however that you understand all of the stages in the process before adapting the methodology to suit your situation because all situations are unique.

The 10-stage GIS planning methodology

Stage 1: Consider the strategic purpose
Stage 2: Plan for the planning
Stage 3: Conduct a technology seminar
Stage 4: Describe the information products
Stage 5: Define the system scope
Stage 6: Create a data design
Stage 7: Choose a logical data model
Stage 8: Determine system requirements
Stage 9: Benefit-cost, migration, and risk analysis
Stage 10: Make an implementation plan

The following 10 chapters of this book detail each of the 10 stages. Let's take a quick tour of the 10-stage method.

Stage 1: Consider the strategic purpose (chapter 3)

Start by considering the strategic purpose of the organization within which the system will be developed. What are its goals, objectives, and mandates?

This stage of the planning ensures that the process and the final system fit within the organizational context and truly support the strategic objectives of the organization. This stage also allows you to assess how information created by the GIS will impact the business strategy of the organization.

Stage 2: Plan for the planning (chapter 4)

GIS planning should not be taken lightly. Forget about actually implementing a GIS for the moment. Just planning a GIS takes a commitment of resources and people. Before you begin, you need the commitment from your organization that it understands the distinction between planning and implementing and that it is prepared to provide the resources needed to make the planning happen. Making the case means understanding what needs to be done and what it will take to get it done. The end result of this stage is a project proposal that makes that case and explicitly seeks approval to launch the formal planning process.

Commitment to the planning process is essential to a successful GIS implementation, especially in municipal government agencies and other bureaucratic public-sector organizations. The project proposal helps to secure the political commitment to the planning process. This is the moment to introduce the GIS planning process to the most senior executives of your organization. Arrange to keep them fully informed of the planning progress. If you receive approval for your planning project and a commitment of resources at this point, your chances of having a successful GIS are high.

Stage 3: Conduct a technology seminar (chapter 5)

Once you have the approval of your project plan, the in-house GIS planning team can be activated. Defining the specific GIS requirements is the primary task in the planning process. You must meet with your customers or clients (those who will use the system or use the output from the system) to begin gathering specifics about the actual requirements from the user's perspective. A highly effective method of soliciting input on the needs of your organization is to hold one or more in-house technology seminars.

In addition to its information-gathering purpose, the technology seminar is also an ideal opportunity for you to explain to key personnel the nature of GIS, its potential benefits, and the planning process itself. By involving stakeholders at this early stage, you help to ensure participation in the subsequent planning work ahead, so that all participants will appreciate the scope of the planning process. The technology seminar is also the place where initial identification of information products begins.

Stage 4: Describe the information products (chapter 6)

Knowing what you want to get out of your GIS is the key to a successful implementation. The "stuff" you get from your GIS can be described in the form of "information products." This stage must be carefully undertaken. You'll need to talk to the users about what their job involves and what information they need to perform their tasks. Ultimately you need to determine things like how the information products should be made, how frequently, what data is needed to make them, how much error can be tolerated, and the benefits of the new information produced.

This stage should result in a document that includes a description of all the information products that can be reasonably forseen, together with details of the data and functions required to produce these products.

Stage 5: Define the system scope (chapter 7)

Once the information products have been described, you can begin to define the scope of the entire system. This involves determining what data to acquire, when it will be needed, and the data volumes that need to be handled. This includes assessing the probable timing of the production of the information products. It may become clear that it will be possible to use one input data source to generate more than one information product, and this can now be built into your development program. Each refinement helps clarify your needs and increases your chance of success.

Stage 6: Create a data design (chapter 8)

In GIS, data is a major factor because spatial data is a relatively complicated thing. In the conceptual system design phase of the planning process, you review the requirements identified in the earlier stages and use them to begin developing a database design.

Stage 7: Choose a logical data model (chapter 9)

A logical data model describes those parts of the real world that concern your organization. The database may be simple or complex, but must fit together in a logical manner so that you can easily retrieve the data you need and efficiently carry out the analysis tasks required.

There are several options available for your system's database design. The advantages and disadvantages of each approach are covered in this section. You should also consider data accuracy, update requirements, error tolerance, and data standards at this stage, as these issues will affect system design.

Stage 8: Determine system requirements (chapter 10)

The system requirements stage is where you examine the system functions and user interface that are needed, along with the interface, communications, hardware, and software requirements. This should be the first time in the planning process that you examine software and hardware products.

By reviewing the information product descriptions, you will summarize the functions needed to produce them. Issues of interface design, effective communications (particularly in distributed systems), and appropriate hardware and software configurations should also be considered during this planning phase.

Stage 9: Benefit-cost, migration, and risk analysis (chapter 11)

Following conceptual system design, you need to work out the best way to implement the system you have designed. This is where you plan how the system will be taken from the planning stage to actual implementation. You may also need to conduct a benefit-cost analysis to make your business case for the system.

Until now, the focus of the planning methodology has been on what you need to put in place to meet your requirements. The focus at this stage switches to how to put the system in place—an acquisition plan. Issues such as institutional interactions, legal matters, existing legacy hardware and software, security, staffing, and training are addressed at this stage.

The acquisition plan that results from this stage of the planning process will contain your implementation strategy and benefit-cost analysis. This plan can be used both to secure funding for your system and as a guide for the actual implementation of the system.

Stage 10: Make an implementation plan (chapter 12)

The final report equips you with all the information you need to implement a successful GIS. It will become your GIS planning book to help you through the implementation process. Developing the final report should be the result of a process of communication between the GIS team and management, so that no parts of the report come as a surprise to anyone.

The report should contain a review of the organization's strategic business objectives, the information requirements study, details of the conceptual system design, recommendations for implementation, time planning issues, and funding alternatives.

The purpose of this book is to guide you through those stages in your thinking. It will give senior executives the context for the questions they must ask about GIS in their organization, and it will tell new GIS managers how to answer those questions.

Chapter three

Consider the strategic purpose

"Strategic purpose is the guiding light. The system that gets implemented must be aligned with the purpose of the organization as a whole."

Consider the strategic purpose

It all starts with the organization. To develop an effective GIS, the GIS manager must have a clear understanding of what the agency or company does, that it has a working plan to do it, and that GIS can help, at least in serving part of the mission. Most organizations have strategic plans. If they do not, it is worthwhile to examine the mandates and responsibilities of the organization to get a sense of its purposes and objectives. It is easier if the organization has an established strategic plan, if the objectives of senior management are the same as the strategic plan, and there is a commitment to achieve the goals of the strategic plan throughout the organization. This sets the direction for GIS planning.

Equally important is the need to understand the specific business model that is intended to achieve the goals of the strategic plan. This, in effect, sets out for each program area what is defined as successful. The business model provides insights into business sustainability; that revenue will be available for the costs of the business model over time. These establish the framework for analysis of the information needed by the organization and for GIS benefit-cost analysis. For sustained support of GIS development, the technology must be recognized as an important component of achieving business success.

Understanding what the organization does and its vision for the future allows the GIS manager to design information products that are directly related to those objectives and hence are valuable to the organization, minimizing the risk of wasting time on planning that is peripheral to the organization's needs.

Organizations adopt GIS on the assumption that it will make their work easier to do, cheaper to do, or better for the customer or constituency. Figuring out what to implement starts with understanding the very purpose of the organization so that what gets implemented is aligned with the purpose of the organization as a whole.

How do you find out what an organization needs? As a GIS planner, you go into the organization and examine the strategic business plan. If the organization doesn't have a strategic business plan, it may take some work to find out what their real business is. But even a relatively detailed business plan tells only part of the story you're trying to uncover. To really understand the business of the organization, you must also undertake an analysis of the mandates and responsibilities of all the major divisions that will be involved within the organization. This requires active engagement of the stakeholders. Eventually the organization will start to reveal its secrets, its objectives, how its business works and what makes it tick. You need to know all of this to plan for an effective GIS.

Most organizations have a strategic business plan. It will consist of some or all of the following components:

Mission statement: Describes the purpose of your organization.

Guiding principles: Outlines the behavior of your organization as it carries out its mission. For example, guiding principles in a customer-driven organization might be "user friendly," "collaborative," "responsive," or "providing better service."

Goal statement: What the organization hopes to accomplish, in general, over a given amount of time. For example, your organization could determine that over the next five years it wants to automate all the business processes in three of its departments.

Program direction: The current direction or strategy of the efforts. The sum of the efforts of the different programs should achieve the overall organizational goals.

Employee development and support: The plan for providing employee training and staff development.

Public interaction: Your public, or constituency's, defined involvement in the development and updating of the strategic plan. This can be done by directly involving them or through indirect methods such as surveys or focus groups.

All of this information, in any form you can find it, will contribute to your overall understanding of the company's strategic direction and purpose. Be creative and focus some serious attention on making sure you know what the organization is all about.

Equally important is a thorough analysis of the mandates and responsibilities of the functional divisions within the organization.

You should go into individual departments seeking answers to these questions:

- How do the individuals who make decisions currently do it?
- What do they need to know to perform their tasks?
- What information products are appropriate for these tasks?

Imagine the following scenario: you go into a government department of forestry to talk to the staff, a forester there.

"What do you do here?" you ask.

"My job is to manage the harvesting of timber in the best interests of the people of the state," she replies.

OK, that's how she sees her job. Next you ask, "What do you need to know to do that?"

She talks around the subject at length, listing many of the data sets she uses, before she finally gets to something tangible. She tells you that what she needs to know is which trees to cut down, when to cut them down, and how to cut them down.

In this simple statement is contained the crux of what information you needed to know, but you had to ask probing questions and be a good listener to filter all the messages and clarify what she really does. Now that you've identified what is needed in order to do this, then you can start to identify some information products that she will need on her desk, information products that will actually tell her which trees to cut down, when, and how, based on the methodology she already uses.

It will take several conversations like this to get a handle on the information that individuals need to carry out their jobs, fulfill their mandates and responsibilities, or to accomplish the objectives of the strategic business plan. The questions you should ask are:

- What is your job responsible for?
- What do you have to achieve?
- What do you have to produce?
- What do you need to know to carry out your responsibilities?
- What information can GIS produce and put on your desk that will provide what you need to know or help you to monitor or keep track of your responsibilities?

Armed with answers to these questions from a range of individuals, you will be able to pull together a picture of the strategic direction of the organization and will begin to understand the information people need in order to succeed. The answers given by management may differ significantly from those of the rank-and-file, so you need to "blend" the answers in order to get a clear picture of the information needed from the GIS.

Asking the same questions in different departments will help generate a picture of the workflows within departments and how they interact with other workflows. Ultimately a comprehensive overview of the organization's processes will emerge.

With this insight the GIS planner can then begin to examine:

- What data is available now that can be used to create the information products needed?
- Where is this data?
- What new data is required?
- What data-handling functions are necessary to turn the data available into the required information products?

When you know what functions are needed and what data these functions most operate upon, you have enough information to define all of the technological requirements (including hardware and software).

The link you establish between strategic objectives, information, and data gives you an audit trail. You identify how information fits into the organization, and the benefits of creating new information. This is the start of a benefit-cost analysis for GIS: what is the benefit of creating an information product for the organization, and what is the cost of producing that product?

Once the information products are available, they will help the organization to fulfill its mandates and responsibilities—to meet its objectives and progress in the direction it wishes. Even better, the new information products may allow the organization to change its strategic direction or even identify new markets and opportunities. This moment, when an organization's leaders finally grasps the full strategic implications of their GIS, is a sweet moment indeed for the GIS planner.

Plan for the planning

"Since the GIS planning process will take time and resources, you need to get an approval and commitment at the front end."

Plan for the planning

A planning proposal document is a useful tool for making the case to management for resources to complete the GIS planning process. Develop this proposal early to help gain commitment from your organization. This proposal should explain exactly what GIS planning will involve and what resources will be needed. The process of gaining support should be taken seriously—commitment is essential—and you should move through whatever steps are necessary to get the appropriate approval and the resources.

Because GIS planning is usually not a short-term effort, it typically represents a serious commitment of time and money. It can take from six months to a year to look at the overall needs of a large organization. That's why you need specific permission to plan.

The proportion of funds needed in the planning stage relative to the overall budget have shifted due to (positive) changes in the economics of computing. In the early days of GIS, when the massive hardware installations required to run the systems could cost over a million dollars, organizations looking at GIS could easily justify spending significant time and resources on planning to make sure they were going about things wisely. Ten percent of the total system cost was an entirely reasonable amount to allocate for planning. But in today's computing environment, with adequate hardware available in the thousands not millions of dollars, 10 percent of the hardware budget would amount to only a few thousand dollars, not an adequate amount for in-depth planning.

The project proposal will include information on anticipated staffing needs, as well as cost estimates for hardware and software. But it will also need to show that the most significant investments will be in data acquisition and development, not in hardware and software. It is possible that the data costs for your GIS will surpass the hardware and software investments once you factor in things like measuring equipment, labor, conversion costs, maintenance, and the licensing of commercial data. As your investments in data grow, so too will the time and resource requirements of ongoing maintenance.

Chapter four

So despite the perennial lowering of costs for hardware and software, the need for thorough and thoughtful GIS planning is as important today as it ever was.

Let's look at a couple of successful project proposals. The first one was developed by an in-house GIS advocate working at a national park, the second by a GIS consultant which lays out a planning proposal for an Australian state government effort. The general ideas about what constitutes a project proposal document can be adapted to the specific needs of your own organization. Comparison of the two efforts show the subtle differences between an in-house effort and the influence of an objective outsider.

At Jasper National Park in Canada a document called a "terms of reference" was created and presented at a meeting with the senior park service management with the purpose of seeking funding for the GIS planning process. The proposed terms of reference document submitted to management by the in-house team contained the following sections:

Jasper National Park
Terms of reference

Work outline

1.0 Project description

    1.1 Background

    1.2 Objectives of the project

    1.3 Project deliverables

2.0 User needs analysis

    2.1 Situational assessment

    2.2 Client base

    2.3 Business requirements and information products

    2.4 Data requirements

    2.5 Technology requirements

3.0 Software/hardware assessment

4.0 Database requirements

5.0 Implementation plan

6.0 Schedule of payment

Contract conditions

    Parks Canada responsibility

    Timing and duration

    Proposal guidelines

Appendix 1   Documents for review

Appendix 2   Ecosystem database inventory

Appendix 3   Client list

    1. Jasper National Park clients

    2. Corporate clients

    3. Tri-council research

    4. Agency clients

    5. Private industry clients and general public

    6. Information management

List of bidders for GIS user needs

Chapter four

The next example shows a successful proposal submitted by Tomlinson Associates Ltd. to the state government of Victoria, Australia, for involvement in a large-scale GIS planning project. Bear in mind that this is a document produced by consultants aiming to get work assisting the GIS planning process.

At this point it is imperative that the project proposal be reviewed, approved, and the resources committed. This is not just money for hiring a consultant; it also includes a commitment of in-house staff resources. A consultant should not entirely do your GIS planning—you must assume that responsibility. Consultants can help, but you must be the one who drives the process, works with it, writes reports, and makes trade-off decisions. When the consultant leaves, you will be left with the system to implement.

Since the GIS planning process will take time and resources, you need to get an approval and commitment at the front end. This will ensure that the planning process will be carried out. If you do receive approval and a commitment of resources at this point, your chances of having a successful GIS are greatly improved. This approval needs to come from the top—from senior management. Sometimes you might have to start with middle management and work through the chain of command but eventually you will have to sell the idea to senior management. Make sure that management signs off on your proposal to plan for a GIS. Once commitment has been secured, you are ready to prepare to assess the needs and requirements of your organization.

Strategic framework for GIS development

Introduction

Project team

Outline of the proposed methodology
    Staff seminar
    Information product definition
    Benefits
    Data requirements
    Error tolerance analysis
    Information priority
    Functional requirement

Benefit-cost analysis

Implementation planning

Request for tender design (technical contents)

Strategy report

Project deliverables

Work organization

Procurement process (optional)

Project timing

Involvement of government staff

Administrative and cost notes

Cost assumptions

Cost summary
    GIS planning project
    Cost break down—by fiscal year
    Cost break down—by primary year
    Optional cost—electoral office procurement process

Appendix 1:   Background and experience of Tomlinson
              Associates Ltd.

Appendix 2:   Curriculum vitae of key personnel

Conduct a technology seminar

"Think of the technology seminar as a sort of initial 'town-hall meeting' between the GIS planning team, the various staff, and other stakeholders in the organization."

Conduct a technology seminar

Now that you have commitment to the planning process, you must prepare to evaluate your organization's requirements in more detail. You'll first need to establish who should be involved and brief all the participants on their roles and responsibilities. After assembling the GIS team, you host one or more "technology seminars" at which everyone comes to understand the planning process and the fundamental concepts and terminology of GIS. Think of these as sort of town hall meetings. At the technology seminar, the in-house team shares the vision of GIS and reviews the planning process. The deliverable for this meeting will be an initial list of information products.

Your planning project needs to be conducted by an in-house GIS team. In a small organization the team could be just one person, but it is more effective to have a small group consisting of a team leader and one person from each department that will need information from the GIS. The individuals from these departments will assist by clarifying the needs of those departments.

Once the in-house team has been established, you can meet with department heads, but even earlier, you should obtain a memo from the CEO or director of your organization that stresses the importance to the organization of the GIS planning process. This memo should request the support of all department heads. As you meet with the department heads make it clear that some of their personnel will be involved in the planning and the fact that they may need to spend several days in the process. This must be stated upfront so that staff are given time away from their normal duties to help with the planning with full support of their manager. The department heads will probably want to know: What is the role of the department in the planning process? What time commitments will be necessary? Who will be needed?

If all you're undertaking is a single-purpose project within your own department, you may only need to visit one department manager, your own, to get the green light. For a project of any cross-departmental scope, you should definitely visit all the department heads separately to secure their support. Find out what is needed for them to reach a level of comfort with your GIS endeavors. Even if you have only one department head to see, and you are undertaking the planning alone, it is important to get support and stress that the planning will take some time, and that you will not be able to do as many other things during this period.

The following heuristics (rules of thumb) regarding resource requirements are based on the experiences of the author in planning a number of GIS implementations at a variety of regional and municipal organizations:

- From start to finish, the GIS planning process in a typical region, government, department, or small municipality takes four to eight months, longer for complex or large organizations.
- The aggregate total of man hours amounts to six to seven person months, more time for complex or large organizations.
- The leader of the in-house GIS team is committed for 70 percent of his or her working time (and would probably opt for more if other duties didn't intrude).
- Departmental staff involved will spend two to six days in the development of each information product description.
- All levels of staff will be impacted in some way and are thus indirectly involved with the study.

A small team of people (ideally with GIS planning experience) should guide the planning process with a consistency in approach and method. The "guiding team" should be made up of a team leader and two other people, one of whom should be from the permanent staff of the agency concerned. The second person on the team should have participated in at least one previous GIS planning study and be fully conversant with the methods and techniques employed. The third person (where there is a team of only three) should be from the permanent staff of the region, city, or municipality and is the one appointed to be the local organizer.

Once you have your in-house GIS team, and the support of department heads for their staff to be involved, you can organize a meeting and call it a "technology seminar." This is a training event that should raise awareness of GIS, outline the planning process, and explain the roles of those involved. One approach is to run an event over two or more days for all the people in the organization who expect to get information out of the GIS. In a large organization there may be 30 or more participants, in smaller organizations it may be just a dozen or less. The agenda for the event should include:

- What is GIS?
- GIS terminology
- Functions of a GIS
- The planning process: steps and responsibilities
- Preliminary/first identification of information products

Successful planning requires effective communication between all those involved in the process. Spend some time during the seminar explaining the nature of GIS, some of the basic terminology, and the functions that systems can perform. This will help everyone in the organization establish a shared vision of GIS and shared language so that they may better explain their needs and requirements.

It is also important to spend some time explaining the planning process and the stages to follow so that participants will know when and how they will be involved and what is expected of them. Since one of the priorities of the seminar is to identify the information products you will generate from your system, you must explain this in no uncertain terms. At the technology seminar it will be possible and in fact necessary to begin identifying the information products that the participants will require.

The primary purpose of their involvement is for them to communicate what they want out of the system (or at least what they think they want from the system).

During the technology seminar you can begin to capture what the participants are saying. Use a whiteboard and assign someone to take good notes. Ask people directly: what's the job you do and what information do you need on your desk? They will probably think about it and say that they don't know. If you prompt them eventually they might give you some ideas. A forester with a timber inventory to manage might say that he could use a better harvesting map, an urban engineer tracking sewer repairs might say that she needs an accurate inventory of sewer hook-ups, updated automatically whenever completed work orders are submitted back to engineering. Some people might get hung up trying to think of map products under the assumption that a GIS is a mapping system and a mapping system only. Remind them that sometimes the best GIS output is in the form of lists derived from spatial analysis like all the properties in a flood zone or customers in a trade area. This means that the new information derived would be in the form of a list (i.e., a list of addresses that must be contacted in an emergency). While these lists can be subsequently mapped, their real value comes from a spatial analysis, not a map. Because you're in a group

Chapter five

session involved in collaborative brainstorming, these first ideas will trigger others in the group to describe the workflow story in their own context. You then turn the thing around and put what you're hearing back into the form of an information product to get consensus in the naming of the information products. By the end of the technology seminar you may expect to collect 10, 20, even 50 information product ideas. It doesn't matter at this point what order the information products will be delivered, some may never get built, but at least you're starting from the largest universe of possibilities you can muster. This process also leads to a great deal of cross-fertilization of ideas, priorities, and cooperation.

At this early stage all that is necessary is the title of the proposed information product and the name of the person who wants it. That person's name is very important and should be required. You need to be able to go back to the person for more details later on and it also provides an indication that someone wants the information product badly enough to be willing to support it themselves. There is no point in defining information products that no one really wants. It is not a good idea to let the head of the department say that he or she must design all of the information products. This person's viewpoint acts as a filter, and at this early stage we want to get the ideas coming in unfiltered. It is the person who is going to actually use the product who really knows what is required and they should be willing to associate their name with the product. Make it a rule: no name, no information product.

Given this initial list of information products you should now perform what is known as a benefit scan. A benefit scan is a somewhat bureaucratic term for an overall look. You should ask how the information products identified would benefit the organization. Think back to the strategic business plan. How do the information products fit into this? Most importantly which budgets will the product benefit? The list should be reviewed with the GIS team and management representatives to identify the most important information products. Check off the most important ones in terms of the value of the benefits that could be derived. Try to rank them from most-to-least important or beneficial. This preliminary ranked list of the most beneficial information products will help in the next stage of planning. With it you'll select the most important and mission-critical ones off the list that you know with some certainty that you can technically achieve, and for these you will draft fully developed descriptions for your first round of information products. A more detailed account of these things we'll call information product descriptions, or IPDs, is presented in chapter 6.

Consider the following program outline that could be distributed to participants in a technology seminar for a typical city, region, or municipality.

The technology seminar is the first face-to-face contact between the GIS planning team and the municipal staff involved in the planning study. The purpose of the seminar is to:

- introduce GIS to the participants (if this is necessary)
- introduce the planning process to the participants
- explain to participants the reason why the work is being done
- make clear the nature of the contribution required of participants, and how their efforts will improve the overall chances for success
- introduce GIS terminology and methods that will be referred to during the study and through final implementation

An underlying objective of the seminar should be to achieve good working relationships between the planning team and the study participants. The planning team should demonstrate that they:

- are competent in the subject area
- have experience in related GIS planning efforts elsewhere
- will strive to get participant's views on information needs and management practice and are not proposing their own views
- can be trusted to aid the region, city, or municipality in clarifying and describing its own requirements

GIS means change, and any time you introduce change into organizations there will be internal debate and usually at least some resistance. Building trust and alliances with your colleagues around the organization who will be affected by your efforts is an important dimension of achieving success. You'll need a set of well-written (and carefully proofread) written documentation to distribute to all participants prior to the seminar. This package should include any relevant memoranda authorizing the study, the agenda of the first meeting, and the timetable for the study. You should also provide some introductory GIS information that people can study before coming to the meeting. You could photocopy articles or sections of textbooks (with the publisher's permission) or even purchase bulk copies of a certain book if you find one you really like. (A list of some books appropriate for this purpose is included in the further reading section in this book. You'll discover at the meeting that some people will have read your document from front-to-back, some not at all, and most will have skimmed it with varying degrees of interest. Look for the people who really take to the introduction to GIS materials; these are people who, like you, have recognized something inherently interesting about GIS, and can thus be recruited as allies in your GIS campaign.

The number of staff organization-wide who'll be invited to a single seminar should be limited to 25 to 30 people, with the department manager making the selection. Smaller GIS efforts in smaller agencies obviously require smaller teams. Among the required attendees are the senior administrative officer and the heads of all departments involved. Their attendance alone, even if they are figureheads, sends a signal that senior management is committed to the effort and should stimulate interest in all invitees to attend. From the departments, experienced staff members with intimate knowledge of their own department processes should attend, as well as any representatives of other agencies such as neighboring regions or provinces with GIS experience. If there are any GIS efforts already established at the organization, then the people involved with that effort should also be included.

The meeting location will usually be limited by available space, but the most satisfactory arrangements are for a large room with comfortable chairs located some distance from the office. Ancillary rooms nearby will allow the group to break up into smaller working units. The main room should have several flip charts or whiteboards, ample marking pens, an overhead projector and screen, computer projection facilities if desired, and one or two large tables. The smaller rooms will require flip charts or whiteboards. Local hotels, universities, and conference centers typically make for safe, easy-to-find locations with appropriate meeting facilities.

Program

The typical meeting would be two days in length. The first part of the meeting will take approximately one day. It may vary in sequence, but should include the following agenda items:

- Welcome and statement of commitment to the project by the senior administrative officer.
- Overview of current GIS status in the agency (including current GIS procurement, GIS activities, contribution of this study) by region, city, or municipal local organizer.
- Introduction to GIS by team leader to establish basic GIS concepts and common terminology. This might include: a definition of GIS, the parts of a GIS, and the functions of a GIS. This could take up a full day alone.
- Explanation of the needs assessment process and reassurance from the team leader that this contribution of participants is central to the work, and that the information product descriptions that they provide are a key goal of the meeting.
- Brief overview and examples of information product descriptions.

Expect many questions and issues surrounding the descriptions of information products. Be prepared to explain the generic GIS functions in detail and the difference between data and information products. Show how interim products will evolve and how information products used in one area might find application in other areas. Encourage people to look for these cross-department applications.

After the initial discussions, an **assessment of information needs** should be begun. This amounts to brainstorming. Encourage open-mindedness. At this stage, a brief identification of the information products is all that is needed: descriptive title, scope, and name of individual who needs it. The key to success in this step is to get participants to think about their overall information needs, to come up with logical names for the information products that they want to have, and to identify the name of the person who needs the product on his or her desk to perform his or her work.

One team member should act as reporter and fill in an "information products needed" form as each new idea is brought into the discussion.

Start with one department head and ask him to state broadly the responsibilities of the department. Write a concise version of this on the flip chart or whiteboard. Then ask other members of the department about their responsibilities. Ask them to identify the types of tasks they are responsible for, the types of decisions they have to make, the need for information at their workplace, and the conditions they regularly monitor in order to do their jobs. Capture the essence of this information on the flip chart.

Now go back and start over, this time asking these same staff to identify a single information product that they could see as being useful. Discuss the product only until you have a good idea of what it contains, making sure that it is an information product and not a data set. Decide upon a brief descriptive title and write this on the flip chart. Add the scope of the proposed product, i.e., who else in the organization will require it. Identify the name of the individual who needs the product, but only if that person is prepared to define it more clearly in the next stage of the planning. If nobody is willing to "own" the proposed idea, drop it from the consideration. This will prevent your list from swelling with mere idle thoughts and half-baked ideas. Identify a second information product from the same department and continue in this fashion until that department is out of ideas.

Now move on to a different department. Repeat the process until there is at least one information product identified for each department and the participants are familiar with the process. Once participants are comfortable with the process, you may wish to split into separate rooms to allow each group to work further on the initial list of information products for that department.

At the end of the day you should bring the whole group together to clarify any problems, and to collate a list of the information products needed. This will allow you to check for duplication. What follows is an actual list of information products developed at an internal technology seminar. Even without any more detailed information beyond these product titles, most of the needs are fairly self-explanatory.

Business workflow

Workflows are models of complex business processes used to gain more efficient operations within the organization. Typical complex business processes include things like land-use development approvals, building loan approvals, permitting, conservation land planning, power outage response, service delivery, and distributed facilities management. All of these are complex business processes which have a workflow associated with them, workflows that can be developed using one of several commercial tools.

Examples of information products (by department)

Engineering and public works

Public works program map and list

Sanitary sewer analysis

Road needs map and list

Pedestrian system needs

Annual sewer needs map and list

Basement flooding situation

Sidewalk ramping needs

Minor hard services needs

Detailed engineering design (CAD)

Storm sewer analysis

Maintenance analysis map and lists

Route optimization map and lists

Sidewalk snow removal analysis

Sidewalk route analysis

Snow removal scheduling

Lane map and public notification lists

Fleet tracking

Waste receptacles/litter analysis

Complaints analysis

Park site maintenance analysis

Stop signal location analysis

Traffic/parking changes

Parking needs analysis

Traffic counts

Housing and property

Legal surveys index query

Select area plan production

Legal surveys drafting (CAD)

Dwelling unit analysis

Social housing acquisition analysis

Housing protection analysis

Housing potential map

Residential loans and complaints map and list

Existing housing and demographics map and list

City property map list

Development activity map and list

Social housing map and list

Amenity map for selected area

Waiting list analysis

Architect/landscape design draw (CAD)

Architect/landscape tech. draw (CAD)

Interior design drawings (CAD)

Streetscape design drawings (CAD)

Economic development

Development projects map and list

City clerk's office

Ward profile analysis

Citywide election data map and list

Ward/poll/voter election data map and list

Fire department

Fire suppression water supply map

Emergency response building floor plans

Planning department site plans

Fire emergency demographic analysis

Emergency route selection

Planning and development

Site planning existing conditions

Permit parking program map and list

Development agreement analysis

43

Examples of information products (by department) cont.

Zoning analysis

Cash-in-lieu of parking map and list

Street and lane closure analysis

Development information map and list

Circulation address labels and lists

Reservation of special-needs housing locations

Registered special-needs housing location map

Property location map and application cross-reference

Area monitoring map and list

Citywide monitoring map and list

Official plan designations map

Area monitoring map/list

Vacant land assessment

Census tract monitoring map and list

Spatial market analysis

Neighborhood monitoring map and list

Transportation flow density map

3-D simulation of urban section

Perspective views of urban section

Ward monitoring map and list

User-defined area monitoring map and list

Overlay mapping system

Employment analysis map and list

Accident rate analysis maps and lists

Spot speed survey map and list

Development site impact analysis map and list

Pedestrian flow map and list

Intersection operation analysis map and list

Employment/FSI capacity map and list

Office and retail vacancy rates map and list

Summary of development projects map and list

Commercial activity map and list

Capital works project status list

Public land facilities list

Inspection status by district map

Cumulative development activity

Inspector performance activity map and list

Recreation and culture

Recreation site analysis and design

Recreation/cultural site location analysis

Recreational/cultural opportunity analysis

Legal department

Lane and street index map and list

Police department

Special-event planning

Emergency route response map and list

Recent occurrence analysis map and lists

Police response property map and lists

Crime occurrence—best suspect system

Workflows are relevant to GIS planners because the GIS tools will often be required to operate in the context of a larger workflow. The key question here is: What's the business workflow design and is there an opportunity to rework how the organization does business to gain efficiency or to capitalize on a new opportunity? This is where a very high percentage of the gains of benefits to the organization will come from. It's the reduction of time or the number of steps in a process, it's the centralized database instead of replicated databases all over the place. So business workflow is becoming more and more important as time goes on. Of course information products come out of that workflow. The data and the data handling also move through the business workflow process. In the context of the business workflow, it is necessary to distinguish clearly between an information product and an application.

An information product by itself passes through the workflow as data through the GIS to create an information product:

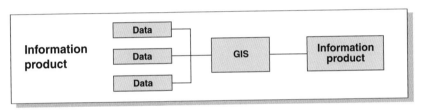

A GIS with data sets embedded within the workflow creates multiple information products. We still talk about information products, but in the context of a business workflow we call the entire bundled thing an application.

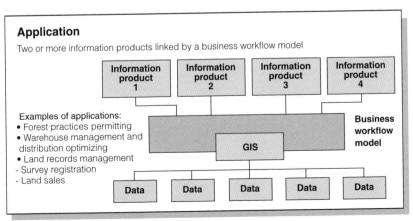

Chapter five

A typical scenario evolves when you begin to press people for what information products they want. What often occurs is that they realize they need to take a hard look at their business processes at the same time. If people are having trouble defining information products, have them go instead to define the actual workflow process and see what information products come out of that.

This is where you ask: Do you want to redesign a workflow in part of your organization? And if the answer to that question is yes, who's going to do it? It may be necessary to contract out the work, which would have to be complete before the GIS could be integrated. This is people thinking about their organization. GIS is an enabling technology which allows them to think about changing their workflow.

The work of including the business flows in the planning process is summarized in the following diagram:

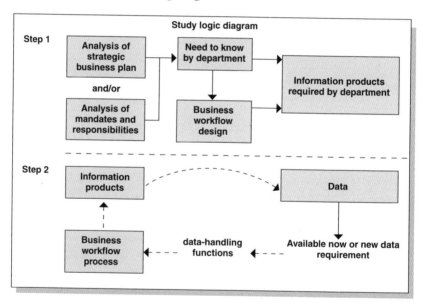

Describe the information products

"Know what you want to get out of it."

Describe the information products

The technology seminar provided you with an initial list of information products and names of the people who requested each product. This is your starting point. The whole unfiltered list defines the potential scope of the expectations that exist in the organization. The next step is to take the most urgent and important of these as defined by your ranking, and start looking at them in more detail. At this point, you will begin to develop thorough descriptions of information—specifications that will allow the actual product to become a working reality.

These documents are the very building blocks of the planning process and we'll refer to them often, so let's use the acronym IPD as shorthand for the information product description.

To create these IPDs, what you're really doing is specifying for the first time what output your GIS must be able to generate. These are the tools you'll leverage to gain approval for spending money on GIS hardware, software, and data. At this stage in the planning process you need to:

- clarify the information products that need to be produced by the system
- establish what data is needed to create the information products
- identify the system functions that will be used to create the information products
- assess the benefit to the organization of having the information product

If this seems like hard work, you are being astute and realistic. This is the mental heavy lifting that is necessary to create well-defined specifications for the information products that you're ultimately going to have to build. "Pay me now or pay me later" warns the old craftman's adage. In some ways, because it is such a creative process, this is also the most interesting part of the process. After this crucial step, what remains of planning your GIS will fall into place systematically.

An IPD contains the requirements for the products that your GIS must be able to create. To be useful, an IPD should include the following components:

- Title.
- Name of the department and person who needs it.

- Narrative summary providing an overview of the information product in layman's terms. One paragraph is usually sufficient.
- Map requirements—a visual sketch or actual example from another source, with a legend.
- Tabular data requirements—details of any information that will be presented in the form of a report, list, or table, including headings and typical data entries.
- Text documents—details of textual information to be included, such as Adobe® .PDF, Microsoft Word, and .txt documents.
- Image requirements—details of images to be displayed in the information product. 3-D output requirements.
- Schematic requirements—examples of type of schematic output required.
- Steps required to make the product—details of the data and software functions needed for just this information product. Identify steps using mobile handheld functions.
- ModelBuilder—allows for wizard-based rapid prototyping of GIS applications.
- Frequency of use—an accounting of how often the product will be used and by how many people each year.
- Logical linkages—details of any linkages that need to be established between data elements in the database.
- Error tolerance—an estimation of acceptable levels of error in the information product.
- Wait and response tolerances—these describe the time requirments.
- Current cost—the costs of producing the product using current methods.
- Benefit analysis—the benefits (or cost savings) of having this information product in your organization.

In particular, map requirements, tabular data requirements, document, image, or schematic requirements help to clarify what output is really needed from the system. Determining output requirements in turn clarifies what must go into the system.

Now let's examine the individual components of an IPD:

Title
An evolved, accurate pithy two-to-three word title developed from the list collected at the technology seminar.

Name of the department and person who needs it
The title, department, and requesting person's name is important. It should be placed on every page of the information product description. This reinforces ownership of each information product.

Summary of the information product
This is an easy-to-understand synopsis of the information product needed and its purpose. This summary is usually the first page of the IPD.

Map requirements
This is the section that describes each map required (if a map is indeed required) in the output. It is important to include some sort of hand-drawn sketch or perhaps an actual example culled from another agency of the desired map. The sketch can be simple, but should show at least one of every feature type that the final product is expected to display. If users need the map at two different scales, this should be indicated as well.

The sketched map or example should include the following information:
- map title
- legend showing how data is presented thematically on the map
- any special symbology required
- colors required
- scale bar
- north arrow

Tabular data requirements
An information product is not always a map. It could also include a list of figures, a table, or a report, or the table or list could comprise the entire information product. This all falls under what is called tabular data. Spreadsheets, databases, and other text files are all part of this broad category called tabular data. This part of the IPD should identify each of

these lists, tables, or reports. Each of these should have a title, appropriate column headings, typical entries, and an indication of the source data file for this data.

If predefined report formats exist, be sure to identify them at this stage. Reports can usually be automatically converted into appropriate formats. Report creation capabilities should be flexible enough to accommodate change in the future.

Text documents

In addition to tabular data, text-based information of a more narrative nature is often an important part of an information product. These include things like Adobe .PDF files, Microsoft Word documents, and plain old reliable .txt files. Be sure to include the likely number of pages per document that will be retrieved from the GIS as well as the database keys that will be used to select the appropriate documents. In addition, you should indicate the need for document display. Specify whether they need to view the document, copy the whole document to hard copy, copy the whole document digitally, copy portions of the document to hard copy, or copy portions of the document digitally. Also specify if changes can be made to any of the documents or images.

Image requirements

An information product may also include an image file or 3-D representations. In fact the entire product could be that image. These days, most agencies hold a wealth of information in the form of image files: digital photographs, satellite imagery, scanned floor plans, and diagrams. In the IPD, you should identify each type of image that the user will need to retrieve from the GIS. You should also include the key identifiers that will be used to search for and find the required images. You should complete this portion of the information product description for every kind of image that users want to retrieve, including scanned documents and video clips.

The document and image requirements can be specified on a single sheet.

The information product descriptions for documents and images should include the number of pages that will be retrieved (typical and maximum), the key identifier that will be used to search for data (an address for example), and what you expect to see on the document.

3-D representation

Three-dimensional (3-D) scene generation capabilities in geographic information systems are not yet true virtual reality, but significant steps are being made toward creating 3-D representation and three-dimensional symbology. There is now the ability to extend databases to include geometry models for textured 3-D solid objects representing buildings and other objects in the landscape. There is the ability to show a 3-D symbology of points, lines, polygons (spheres, cubes, strips, dotted lines, etc.) and realistic out-of-the-box texture patterns. Extensive libraries of 3-D objects that can be used to simulate 3-D space will be available, as well as translators from common 3-D object formats such as Open Flight, 3-D Studio (realistic 3-D models with a CAD orientation), and VRML. Symbology attached to these models can be geospecific in the shape of a 3-D marker symbol, or geotypical in the shape of a feature in a feature class. Continuous multiresolution global viewing of geographic information is becoming possible. This is achieved in ESRI ArcGlobe™ software, which supports the dynamic 3-D viewing of the information working directly from a geodatabase. In short, three-dimensional representation is becoming easier, quicker, and more flexible, and the foundation is being built for future development.

Mobile handheld capabilities

A range of handheld devices and tablet PCs with differing capacities and operating systems are now available to interact with GIS, and they need to be considered during this stage of the process so that you'll know, for example, if an application needs to be available on desktop PCs as well as handheld computers. These devices offer important new capabilities for GIS developers, including the ability to operate through loosely coupled architecture for disconnected editing of the central database in the field, and should not be overlooked.

Schematic requirements

An information product may present data in the form of schematics for ease of understanding. These can approximate the real world using geographic coordinates, can present a geo-schematic emphasizing the topology of a network or other system, or be a pure schematic showing only the flow paths as linkages. A geo-schematic of the water system connecting the fire hydrants may be valuable, for example, in deciding from where to draw water for a large fire response. An example of a schematic representing an electrical network is shown:

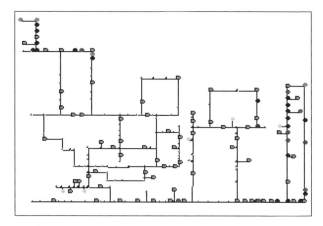

Orthogonal schematic of the selected set (in gray dots)

Overlaid with street network

Steps required to make the product

The last components of the IPD described above (map, list, document, image, and schematic requirements) clarify details of the information product that is required. Once you know something about the information product, you can begin to evaluate the steps needed to make it. These steps identify both the data and functions required to make the product.

As you start to examine some of the data sets you'll be accessing, you'll notice that conflicting names are sometimes used in different departments to refer to the exact same data set. Therefore it's important to carefully study the files and naming conventions to establish one standard name for each data set. It may even be necessary to create a directory of synonyms just to sort things out (more on data dictionaries later). It is good practice to use a single simple descriptive title that is unique to the data set. You will use these titles later to create something called a master input data list.

You should now write a step-by-step description of how to make the information product, a rundown in order of the manual steps (many of which will later be automated) that are required in order for the GIS to spit out the thing you want. It covers all stages from the initial request for the product to completion. In fact, with GIS there are often several ways to do many things, but for now all you need to do is specify one logical and direct method. This clarifies the thinking about the real objectives of the user and starts the process of identifying all the data required. Later on, the system will be fine-tuned to create an application that makes elegant use of the system functions.

Often, the most effective way to identify the sequence of steps needed to make an information product is to think of how you would create the product by hand, using hard copy data sources. Setting down the steps needed in this method is not difficult, but it requires logic and a linear thought process. The key is to stay focused on the end product you're trying to create and not to get distracted by other possibilities inherent in the data or by functions that you don't need. Rigorous care at this stage will often resolve questions about data availability and suitability and identify sources of potential error.

Here are some additional tips that will help make your description easier to write and more useful thereafter:

1. Use one data set at a time and use its standard name. Write down the source scales of any maps that will be used as a reminder of the source data's resolution.

2. Use the generic function descriptions provided in the lexicon. They are easy to understand and will translate into any system-specific piece of software.

3. Every time you use a function, explain the work that is being done on the data set and clearly identify the data elements that have to be extracted in the process. This is very important, as the data set could contain many more data elements than the ones you need and you need to make sure you're working on the right one. You will also cross-check the product requirements against the right-hand column of steps to make sure that all the data elements needed in the product have been produced by the functions working on the data sets.

4. Always operate on the assumption that the results of using any function are available for use in later steps. For more complicated processes, you can represent the steps visually in the form of a flowchart to aid in communication. The steps to make the product should be clear and should help identify the data needed, functions invoked at each step, details of any interim products, and the final product.
5. If data is not readily available, now is a good time to think about how you will obtain it—from the government, a data vendor, or in-house creation.
6. If the staff involved is not aware of the range of GIS functions available or the terminology to describe them, you may need to seek further training in basic GIS functionality.

ModelBuilder

Recent developments make the development of prototype GIS applications far quicker and easier than ever before. ModelBuilder software by ESRI allows wizard-based and graphic-based construction of GIS applications utilizing a wide variety of data types and geoprocessing functions.

ModelBuilder has the advantage of being a rapid prototyping tool that allows the combining of known data sets with a series of geoprocessing function steps to produce information products. These can be simple or complex products; in fact, they can be extremely complex. They can be combined models. The output can be examined as the model goes through several stages of development. ModelBuilder will play an important role in iterative design, as a model can be shared between people who can add to it, revise it, and implement it. Given an accepted model, it can be published, rather in the way an application is currently published, but with the steps in the model graphically illustrated and amenable to revision.

A greatly improved version of ModelBuilder will be released in version 9.0 and will have the advantage of a broad set of new geoprocessing tools. It is possible to visualize advanced versions of ModelBuilder being used in complex simulation studies.

Future directions of development include building interfaces between ModelBuilder and business workflow applications such as Visio® Enterprise, ABC Flowcharter, or Workflow Analyzer, where ModelBuilder can build the information products required by the business workflow, and similarly to interface with CASE tools for the physical design of the databases necessary for the models. In the absence of existing databases,

ModelBuilder could set the parameters for the design requirements of such databases so that they can be established with the ability to make the information products required. This could avoid the crippling result of data-centric GIS development where huge databases have been established which have no ability to produce the information products they must provide.

Frequency of use

After determining all of the functions and when they are used in the process, you can do a useful summary of the number of times each function will be used in the creation of the information product. You may have information products that require multiple functions and those functions may have to be invoked many times.

Related to frequency of function invocation, you also need to estimate the overall demand for the information product. You should estimate how many times per year it will be generated. The frequency may range from once per year (a map to hang on a wall) to 10,000 times per year (some emergency routing products).

You should list the number of maps you anticipate producing, the number of lists you think you will generate, and the number of documents you will retrieve each year. For the IPD, try to project these numbers out to five years.

A simple table should be included in the IPD summarizing the frequency-of-use estimates.

After you determine the number of times per year the product is generated, you can multiply that number by the number of times a specific function is used to make the product. The result is the number of times the function is used in a year to make that product. For each year, the equation is:

Number of times a product is created in one year	X	Number of times a function is used during creation of the product	=	Number of times a function is used to create the product in one year

You are now getting a first view of the overall use of functions in the system you need. After you've created your first round of IPDs, you can create a list of the top-ten functions that will be required from the GIS overall. Such a list will be useful in scoping your system requirements for hardware and help you evaluate your choices among potential GIS software packages. The list is an important tool for ensuring that you obtain a system that effectively and efficiently meets your needs.

The complete process of determining frequency of function use is discussed further below.

Logical linkages

The next step in describing an information product is to determine the relationships that are required between the data elements and data sets. These relationships are called logical linkages, and they must be in place when you build your database later on. In the IPD, you need to establish how data from different data sets will be joined to create the end product.

There are three types of logical linkages:

1. Relationships between lists and graphic entities: these are relationships between features (points, lines, polygons) and their characteristics (attributes). (i.e., names attached to items)
2. Relationships between maps or map layers: these are relationships between the different kinds of maps (or data layers) that you require. (i.e., can they be overlaid; same scale, map projection)
3. Relationships between attributes: these are relationships between characteristics and between data elements. (i.e., does this link to that thing)

Don't look for every linkage you can establish, but do look for linkages that might not be there. If you can identify logical linkages that are necessary but not currently in the data, you have identified a problem that must be solved before this information product can be made.

If, for example, you know that you need a linkage between a house and a sewer, you can input the information while you're building your database. If you find out after you've built your database that you need a linkage between a house and sewer, you might have to revisit every house and sewer to input the data. This might not be significant if you have a town of only 500 dwellings, but if you have 120,000 dwellings to revisit, the problem becomes significant indeed. Information about linkages becomes more important as the size of your database increases.

Error tolerance

Error tolerance is the amount of error that you can accept. The question you must answer is, "How wrong can it be?" When someone requests an information product, it's because they expect it to provide a benefit. It might save staff time or increase operation efficiency. If the information product contains errors that require staff time to correct, or that require another method for determining the right answer, you receive neither the savings in staff time nor the improvements in efficiency.

The information product description should address possible errors, the result of errors, the impact on benefits, and the amount of error that is acceptable. That is, you must establish how much error can be tolerated in the information product and still have it be useful. It is important for users to think about error and to understand what it means to them in terms of cost versus reliability.

This error discussion represents the user's view of their needs. This is their "bargaining position" at the start of the discussions on data quality during implementation. Once costs and benefits are weighed, the user may not get their desired accuracy, but they will be prepared and will understand the subsequent reliability of information products.

The levels of quality control established to ensure that data meets user needs can significantly affect costs. Achieving the elusive final 10 percent of accuracy may cost 90 percent of the total costs of building the database. With users, you should determine the maximum amount of error that can be tolerated while still providing the benefits intended from the information product. You need to find out where the balance is between accuracy requirements and data-entry and quality-assurance costs.

There are four types of error:

- Referential: an error in a reference to something, such as a wrong address, label, number, or name
- Topological: a linkage error in spatial data, such as unclosed polygons or breaks in networks
- Relative: an error in the position of two objects relative to each other
- Absolute: an error in the true position in the world of something

Error tolerance is expressed numerically, but differently depending on the type of error. Referential and topological errors are usually expressed as a percentage of the total error, while relative and absolute errors are

usually expressed as a linear distance. Below are examples of error tolerance expressions. Consider the example of how each of the four types of
error might be expressed.

Error type	Error	Tolerance
Referential	Incorrect street address	2%
Topological	Incorrect street network	0%
Relative	Sewer line shown on the wrong side of the street	± 2 m
Absolute	Floodplain not aligned with property boundaries	± 10 m

The following tables show examples of each type of error.

Referential error

Possible occurrence	Incorrect street address.
Results of error	Delivery to wrong property.
Impact on benefits	Time wasted during delivery reduces benefits of having product. Pizza may get cold.
Concerns for error tolerance	Must balance data entry and checking costs with amount of acceptable error, e.g., 2 percent of addresses wrong.

Topological error

Possible occurrence	Incorrect street network.
Results of error	Route may take longer than necessary.
Impact on benefits	Loss of life or property because an emergency services vehicle is delayed.
Concerns for error tolerance	If loss of life may occur, zero tolerance may be required.

Relative error

Possible occurrence	Sewer line shown on the wrong side of the street.
Results of error	Excavation for sewer repairs in the wrong place.
Impact on benefits	Increase in on-site costs.
Concerns for error tolerance	How much error can the sewer line have before it's ineffective? It might still be effective if it is only a little off-center, but if it is on the wrong side of the road, it's ineffective. ±6' (2m)

Absolute error

Possible occurrence	Floodplain boundary not aligned with property boundaries.
Results of error	Uncertainty of location. Are you in the floodplain?
Impact on benefits	Owners paying for flood insurance when it is not needed. Owners not being protected when they are in a floodplain.
Concerns for error tolerance	Need to balance costs of data, costs of checking, and the amount of acceptable error. ±30' (10m)

In thinking about data remember that there is no such thing as inherently "good" data. In working terms there is useful data and useless data. Data also has no intrinsic value or accuracy. There is no accuracy that data should have—only the accuracy demanded by the information to be created from the data that impacts people doing real work. Data is useless if the accuracy is not related to the reality of what the data is being used for.

Wait and response tolerances

Wait and response tolerances are related but distinct concepts that must be included in the definition of the information product.

Wait tolerance is a measure of how robust the computer and network system must be, i.e., what is the maximum allowable time (with the computer up-and-running) between the last keystroke and the full display or output of the information product. For some emergency dispatch uses, the error tolerance can be as low as seconds, for other applications it can be irrelevant. This has an impact on network design. If they apply to the information product, they could be anywhere from extra low (0 to 1 seconds) up to very high tolerance (up to an hour).

Response tolerance has to do with process. This is the maximum time between the arrival of the request or critical data at the GIS office and the delivery of the information product to the user. This will help you understand how many GIS office hours, staff members, etc., will be needed to meet the demand. For example, what happens when the forest burns on Sunday and the maps aren't available until Monday?

Current cost

The next step is to identify what it currently costs you to create the information product without GIS. Your estimate should include both labor and materials. It is also important to identify how many times a year the total cost occurs.

The cost figures you calculate can be used in cost-cost comparisons and benefit-cost calculations to help justify the implementation of the GIS. For example, if an information product costs a few hundred dollars and is used only once or twice a year, the automation costs will not be recovered in a reasonable amount of time. On the other hand, if you have an information product that takes a lot of staff time to create but is needed frequently, the current cost will probably justify automation. This step helps you identify the level of effort required for each information product being requested.

It is useful to remember that some of the data characteristics that you define may increase costs. For example, the more precise the data needs to be, the more it will cost. Although many projects require high accuracy and precision (e.g., gas and water utility management and engineering applications), some do not (e.g., trade area and census demographic analysis).

Benefit analysis

The final step in preparing the IPD is to perform a benefit analysis. You should perform a benefit analysis whether it has been explicitly required or not. As a GIS manager, you should know the benefit of the information you propose to create from the GIS. You should be able to weigh the costs of system and data acquisition against the benefits that your organization intends to receive from the information products created.

A simple model examines the following three categories of benefits:

- Financial savings: actual cash saved from current budgets if required information products were made by the GIS (e.g., reductions in current staff time, increases in revenue).
- Direct benefits to agency: things that will result from new information that was not available prior to implementing the GIS. These could include improvements in operational efficiency and workflow, or the reduction of a liability.
- External benefits: benefits that accrue to others who are not directly using the GIS. For instance, the general public would benefit from lower fire insurance rates if the fire department had better fire response times due to having immediate access to reliable maps.

Once you've identified the benefits of each information product, you can total the benefits of all information products and compare the benefits with the cost of implementing the GIS system and acquiring the data needed to make the information products.

Finalizing the information product description

When the IPD is complete, it is important to get the person who is requesting the product to review and initial every page and then sign the last page of the IPD. This person must sign every page to confirm that the description accurately describes the information product requested. If there is no signature, there should be no information product. It is best to make this clear from the beginning, so the person knows that they are responsible for making sure the final description meets their needs. People sometimes don't realize the amount of work that goes into building a GIS. It's important that they are fully aware that you're going to be spending precious time on their behalf, and that you want to spend it on information they need.

It is essential that the department head overseeing the budget to which the benefits will accrue initial and sign the benefits pages. If the department head will not sign the benefits pages, find out why not and make the needed revisions. If you can't reasonably accommodate the changes, then there is a fundamental problem with the information product. If there is no department head signature, there should be no information product. This goes back to the recommendations made earlier—make sure you have full support from upper management from the beginning of the project and regularly inform management of progress.

A consistent and rigorous approach to approvals is what makes GIS successful in established organizations.

The value of information

Calculating benefits is always a difficult task, but it is a critical one in the GIS planning process. At one time, most economists professed that in theory the only legitimate value that could be attached to data was the value that someone was prepared to pay for it. In practice this is ludicrous, because no one buys data just for the sake of buying it. People buy data because they want to do something with it that is going to create information that will benefit them. New models are required that consider the benefits that accrue from having the information.

Although calculating the benefits that accrue from having information requires some homework. The author has developed a methodology for the measurement of benefits of information. The methodology has been tested and proved successful in Canada, Australia, and in state governments in the United States. The organizations that used the methodology accepted the outcome and the results from it, which is the real measure of success when performing benefit-cost analysis.

Case study

Here is a more detailed example in the form of an actual case study: creating an information product description.

Context: Gary works in the engineering department of a major North American city. He is charged with the maintenance and repair of the sewer system. He needs an information product that will help his staff deal with frequent sewer backups. He envisions a sewer incident situation map being produced by the GIS and has identified himself as the person requesting this information product during his city's recent technology seminar.

The background is as follows: in this city the sewers are generally known to be in a terrible state. The original lines date from the middle of the nineteenth century. At one time, they were combined sanitary and storm sewers, but were separated later on. Fast-forward to the present and the city's engineering department is dealing with a large frequency of trouble complaints. The aging and modified sewers tend to flood causing people's houses to get filled with backup sewage—a very unpleasant situation to say the least. Every year there are an average of 50 sewer backup incidents all over the city. This is not a trivial situation. When a backup occurs, homeowners call city hall for action and the engineering department must react quickly.

This example examines in detail the IPDs for the **sewer backup situation map** that Gary needs.

Summary

Gary is the person identified on the IPD. He will be responsible for making sure the IPD accurately describes the information product he needs. Once the information product is complete, he will need to sign each page, verifying that the description is correct.

In a nutshell, Gary needs to be able to input the address of the sewer complaint and have the GIS return an information product that includes among other things a view of the property or properties involved, the most probable suspect sewer segments and other segments connected to them (both upstream and downstream), manhole locations, and any other potentially relevant landmarks in the immediately surrounding area. In addition to the map, Gary also needs a report documenting complaint history in the same segment area and the known condition of the segment according to the most recent inspection reports. He also needs to retrieve any available photos of the suspect sewer segments.

Map requirements

What follows is Gary's list of map requirements in support of the sewer incident report:
- Property numbers (addresses)
- Locations of the houses whose sewers are backed up
- Property ID numbers of the affected houses
- Sewer segments the houses are connected to
- Street names
- Manhole numbers

Gary makes a sketch of the type of map he requires.

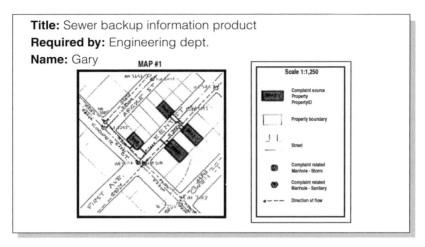

Title: Sewer backup information product
Required by: Engineering dept.
Name: Gary

Sewer backup situation map

Tabular data requirements

There is also some tabular data in the form of three lists that Gary needs as part of the complete information product. The first thing he needs to see is a list that provides the pertinent details about the complaint:

- Address of the affected property
- Property ID number
- Name of the occupant/owner
- Housing type: e.g., single-family residential
- Number of any previous complaints
- Details about the relationship between the sanitary sewer and the storm sewer (a number of previous backups can be traced to this source)

The IPD entry format for this list is seen below. It contains the headings, typical entries, and source information—all the information needed to build the list. The information comes from one of three sources: the complaint file itself; a database of property development information stored by the assessor's office called the PDIS; or the SIMS, the sewer characteristics file maintained in Gary's department on a PC.

Title: Sewer backup information product
Required by: Engineering dept.
Name: Gary

LIST #1

List title: Complaint source

Headings	Address	Prop ID#	Owner occupant	Housing type	Zoning	Previous complaint	Type	Connection	
								Sanitary	Storm
Typical Entries	26 Cooper St.	32968	M.Dewe	SFR	5ac	Oct. 85 Nov. 86	Basement flood Basement flood	To 07062	To 07062
	17 Kent St.	37210	A. Brown	SFR	5ac	---	---	To A3694	To 07062
Source	Complaint file	PDIS	PDIS	PDIS	PDIS	Complaint file	Complaint file	SIMS	SIMS

A second list that Gary needs contains the following information about the suspect sewer segment:

- Segment number
- Street name
- Manhole details
- Sewer capacity
- Sewer size
- Sewer grade
- Construction materials
- Date of installation
- Physical class
- Last inspection date
- Number of floodings
- Dates of floodings
- Reference to the TV report
- Estimated flow

The table on the following page gives the needed specs about information needed for the IPD.

Chapter six

Title: Sewer backup information product
Required by: Engineering dept.
Name: Gary

LIST #2

List title: Suspect sewer report

Headings	Storm or sanitary	Seg-ment #	Manhole		Capa-city	Size	Grade	Material	Installed	Class (Phys. cond.)	Last inspected	Last cleared	No. of floodings	Previous incidents	Water depth during flooding		TV report	Rainfall	Est. flow
			To	From											Manhole	Basement			
Typical entries	Sanitary	A2101 Cooper St.	3694	3695	350 gps	10"	2"	Metal	1936	3	Sept. 86	Sept. 86	5	Oct. 75, Nov. 82, Mar. 87, Jul. 87, May 88	1.5 ft.	1.0 in.	File #123	2"/hr	250 L/hr
	Sanitary	A3694 Kent St.	3694	3697	220 gps	8"	5"	Concrete	1944	4	Oct. 75	Oct. 75	1	None					
Source	SIMS	SIMS	SIMS	SIMS	SIMS	SIMS	SIMS	SIMS	SIMS	SIMS	SIMS	SIMS	SIMS	SIMS	SIMS		SIMS	Inspector	Inspector

The third thing Gary needs in list format is information from the previous management reports on sewer backup incidents, including where incidents occurred, on what dates, and the cause. Gary prepares this table for the IPD:

Title: Sewer backup information product
Required by: Engineering dept.
Name: Gary

LIST #3

List title: Management report

Headings	Sewer section	Flooding dates							Tot. # of flooding
		1	2	3	4	5	6	7	
Typical entries	A2101	Oct. 75	Nov. 82	Mar. 87	Jul. 87	May. 88			5
	03624								
Source		SIMS—Management report							1

Document requirements

After the maps and lists are identified, Gary next specifies any text files he may need to access (for example, a transcript of the inspector's last taped voice report).

Image requirements

The engineering department employs a robotic video system capable of delivering images showing pipe condition in the segment. These images are all date-stamped and referenced to the segment number.

The form specifies that Gary will retrieve these documents based on sewer segment number. He needs to be able to visually observe the documents and copy portions onto hard copy, but not change the documents in any way.

Now you know what Gary wants. He wants maps, lists, text files, and images. Start thinking about how this information product, with its disparate data sources, can be produced:

Title: Sewer backup information product
Required by: Engineering dept.
Name: Gary

Scanned document display			
#	**Data set name:** Sewer characteristics (SIMS) file		
Document title:	Sewer TV reports file		
No. of pages per retrieved document		Typical 2	Maximum 5

Search keys (all)

Spatial: Sewer segment number

Attribute: --

Data elements (Required to be seen) Sewer interior TV scan (manhole to manhole) of suspect sewer segment

Action:	✓	Visually observe	Read only
(Check as appropriate)		Copy whole	Hard copy
		Copy whole	Digital
Change:	✓	Copy part	Hard copy
(Check as appropriate)		Copy part	Digital
		Add data	Which elements
		Delete data	Which elements
		Edit data	Change errors
No change permitted	✓		

- What data do you need to make the information product?
- Where will the data come from?
- Which functions are needed to create the products?
- How do the component parts relate to one another?
- What common data elements in the different data sets are needed to link them together?

Currently, only the property information and a sewer map exist in the city engineering department. Gary needs to contact the water company to get the sewer condition data. The sewer condition data should be linked to the property file to allow sewer segment numbers to be linked to street addresses. Digital sewer data is lacking within the organization and its creation would be beneficial.

Steps to make the product (data and functions required)

The table on page 74 gives details of the step-by-step process required to make Gary's information product. From his description of the process (column 1), the data sets required and the system functions that will be necessary to make the product have been identified and organized into

70

the "steps to make the product" document which should be included in the IPD.

The far-right column in the table on page 74 and 75 is very important. This column contains information that not only describes the functions used, but specifies every data element that must be taken from the data set and included in the information product. This is used later to confirm that all the data elements needed will actually be in the product.

Title: Sewer backup information product
Required by: Engineering dept.
Name: Gary

Function	Number
Data input	1
Attribute query	5
Network analysis	1
Graphic overplot	2
Spatial query	1
Scale change	1
Display	1
Edit	1
Label	1
Symbolize	1
Plot	1
Create list	1

The system functions needed to create the information product (and the number of times they have been used) are listed on the previous table.

Multiplying the number of times each function is used by the number of times per year that the product will be created will give an idea of functional utilization. In turn, this indicates the type of system that would be appropriate.

Gary estimates that 50 of the information product called sewer backup complaint map products are required each year.

On the next page are the number of times each function would be used in a year to produce this information product. The functions have been ranked so that the functions used most often are listed first.

Chapter six

1. Attribute query: 250
2. Graphic overplot: 100
3. Data input: 50
4. Spatial query: 50
5. Network analysis: 50
6. Scale change: 50
7. Display: 50
8. Edit: 50
9. Label: 50
10. Symbolize: 50
11. Plot: 50
12. Create list: 50

The most commonly used function is attribute query. The software that Gary's organization implements must be able to perform attribute queries efficiently for the benefit of this information product.

Logical linkages

Gary needs a link in the GIS database between street address and property boundaries in order to select the lot lines for the map. He also needs a link between a sewer segment number and the sewer network segment, the actual line in the digital database. These are key linkages, between items in a list and a graphic entity in the database.

There are also some map-to-map linkages required. Gary needs to be able to overlay the property boundaries on the sewer network map and the topographic map—meaning they'll need to be available at a common scale and projection. Note that the degree of scale change of any data set required to make the overlay should not be too large. As a rule of thumb, scale changes of over 2.5 times in either direction should be avoided, but you may have to live with more in some cases if the more closely projected or scaled data is not available. Heed the information systems creed: be guided by the error tolerance.

Finally, there are three attribute-to-attribute linkages that must be made: the street address needs to be linked to the property attributes in the property information system (PDIS); the street address needs to be linked to the sewer segment numbers to determine which sewer segment in which sewer is backing up into each house; and the sewer segment number needs to be related to the sewer characteristics file. It is useful to list out all the linkages in one place for the IPD as seen on the table below.

Unfortunately there was no sewer segment map even in existence, so it is not possible to link sewer segment number and sewer characteristics.

Title: Sewer backup information product
Required by: Engineering dept.
Name: Gary

List to graphic entity
  Street address to property boundaries (polygon)
  Sewer segment number to sewer network segment (line)

Map to map
  Ability to graphically overplot property boundaries on
  sewer network map and topographic map

Attribute to attribute
  Street address to property attributes (PDIS)
  Street address to sewer segment number
  Sewer segment number to sewer characteristic file (SIMS)
  attributes

Because there are no sewer segment numbers from a sewer segment map, it is also not possible to link street addresses to sewer segments. To the uninitiated, this may sound like a simple thing to correct: just manually determine and append the data. But in fact the estimated cost of creating such things can be enormous—in the millions of dollars for even a medium-sized town or city—because the only way to do it involves scanning old sewer plans, digitizing the real position of the manhole covers from aerial photography, sliding the images in behind those digital points, and creating a sewer network map.

In Gary's organization, the benefits of this new database had been recognized long ago, and it was also decided that the development of the database should not be a cost in the GIS budget, but rather an infrastructure cost in the engineering department's budget. In your own organization, this may not be the case. You may need to figure out smart ways to get these databases built, or else be prepared to include them in your GIS budget.

Chapter six

Title: Sewer backup information product
Required by: Engineering dept.
Name: Gary

	Gary's description	
Step 1	A staff member receives a complaint phone call and uses the complaint system file to get details of previous complaints of that type at that address.	
Step 2	The staff member matches the complaint address with the owner/occupant name, housing type, and zoning information, using the property development information system on the city's mainframe. A property, residential, or occupant file may need to be accessed.	
Step 3	Now the staff member needs to obtain data about the sewer segment at fault. First, the correct sewer segment based on the street address of the complaint must be found. Then, the characteristics for each suspect sewer segment must be found.	
Step 4	Next, the staff member needs a map of the complaint property that includes the boundaries of the property.	
Step 5	Next, the staff member needs to identify sewer segments within one kilometer that are adjacent to the suspect sewer segment. A map showing upstream and downstream sewers within the one-kilometer region is needed, overplotted with information about the city.	
Step 6	Finally, the staff member needs to add street boundaries to the area of interest. The street boundaries are at a different scale than the rest of the information, so a scale change is necessary. With all of this data, final maps and lists can be created.	

Data needed	Function needed
Complaints system file *(Note: there is tight security on this file)*	**Keyboard input** of address and nature of complaint. **Attribute query** to determine previous complaints (date and type).
Property development information system (PDIS) (City mainframe database) Property file Residential file Occupants file	**Attribute query** on PDIS to match complaint address with owner/occupant name, housing type, zoning.
Sewer characteristics file (SIMS) from land-use maps (1:1,250) Sewer plan and profile sheets Sewer log file Sewer TV reports file Drain card file *Note 1: Sewer data must be linked to a sewer segment number which is linked to a street address* *Note 2: This file does not currently exist in digital format*	**Attribute query** on sewer database to match street address with sewer segment number. For each suspect sewer segment, identify manhole (to and from), capacity size, grade, material, date installed, physical condition, date last inspected, date last cleared, and dates of incidents.
Legal survey map (1:1,250)	**Attribute query** by street adress (matched to property identification number, if necessary) to identify complaint property boundary.
Sewer network map (to be created at 1:1,250 scale in topological network form, with manholes as nodes, sewer segments between manholes)	**Attribute query** to identify suspect sewer segments. **Network analysis** to identify adjacent sewer segments within 1 km of suspect segments. **Graphic overplot** sewer segments and city to identify the area of interest within the city.
Topographic map (1:2,000)	**Spatial query** (by region) defined by sewer segments. **Scale change.** **Graphic overplot** selected street boundaries, selected sewer network, and selected property boundaries. **Display, edit, label, symbolize, plot, create list.**

Chapter six

<table>
<tr><td colspan="5">Title: Sewer backup information product
Required by: Engineering dept.

Name: Gary</td></tr>
<tr>
<th>Type of error</th>
<th>Possible occurrences</th>
<th>Result of error</th>
<th>Impact on benefits</th>
<th>Error tolerance</th>
</tr>
<tr>
<td>Referential</td>
<td>Street address error</td>
<td>Wrong identification of property</td>
<td>Erroneous situation analysis</td>
<td>0% error</td>
</tr>
<tr>
<td></td>
<td>Sewer segment number error</td>
<td>Wrong identification of suspect sewer segment</td>
<td>Wasted time while errors are resolved</td>
<td></td>
</tr>
<tr>
<td>Topological</td>
<td>Link between property and sewer not established</td>
<td>Property not included in analysis</td>
<td>Incomplete situation analysis</td>
<td>0% (complete topology required)</td>
</tr>
<tr>
<td></td>
<td>Break in topology of sewer network</td>
<td>Adjacent sewer segment would not be included in analysis</td>
<td>Potential source of problem may be missed</td>
<td>Links between property and sewer are particulary important</td>
</tr>
<tr>
<td>Relative</td>
<td>Location of sewer line within street</td>
<td>Wrong site for excavation</td>
<td>Increase in on-site cost</td>
<td>± 1m</td>
</tr>
<tr>
<td>Absolute</td>
<td>Graphic over-plot of property boundaries on sewer network may be misaligned</td>
<td>Uncertainty in sewer line position with regard to property</td>
<td>Potential increase in on-site investigation costs</td>
<td>± 1m</td>
</tr>
</table>

Error

Gary had to think carefully about how much error he could tolerate. Was it really necessary to check every address three times to get zero errors? Even simple operations such as rechecking an address could have severe impacts on the cost of creating the information product. The crux of the problem is in the question: "How wrong can you be?" How much error is tolerable? Error and accuracy matter because of the impact they have on the reliability of the system as a whole.

Gary could identify four types of error (see table on page 76). He specified 0 percent error tolerance on referential and topological errors. This is the ideal case. If the costs are too high, he is prepared to revisit his error tolerance. Even if Gary ends up changing his error tolerance, it is still important for him to examine error and discuss it with you in specific terms. By examining acceptable error, he starts to understand the impact of errors on the reliability of his information product.

Wait tolerance and response tolerance
Because the sewer backup represents a time-sensitive issue, but not life threatening, the wait tolerance does not apply. The response tolerance was one day.

Costs and benefits

Gary estimated that using current non-GIS methods takes one hundred hours to create this information product, or $2,237 in labor costs each time the product is created. In addition to labor costs, he adds another $100 in material costs for a total cost per information product of $2,337.

Creating this information product 50 times per year has a total annual cost of $116,850. That is the current cost of creating the product without the GIS.

Title: Sewer backup information product
Required by: Engineering dept.
Name: Gary

	Hours	Cost
Labor Professional Technical	100	$2,237.00
Materials		$100.00
Total cost		$2,337.00

Gary knows that the operational budget for sewer maintenance is $12 million per year. The GIS could possibly create the same product in four to eight hours, compared to one hundred hours of manual effort. The time involved would be reduced by over 90 percent. Some staff members would still be involved, but staff time savings of over 80 percent would be realistic. This represents a real savings, a cash benefit.

In addition, the timing of the creation of the product would be improved by the use of GIS—the product could be available within a day as opposed to perhaps three weeks later. This is a benefit to the organization. It will help them react faster to a flooding situation. They may be able to apply solutions in time to make quick corrections possible and hence reduce the anger of the homeowner. Improved timing may also make it possible to coordinate repairs with maintenance operations already underway. If they can't save $100,000 per year from a $12 million per year annual budget, they aren't really trying.

Another benefit to the agency will be a reduction in liability. Homeowners are irritated by long delays in solving backup problems. Gary knows there is currently a backlog of 10 court cases involving unresolved sewer backup problems, including a class action suit for $600,000 by a group of homeowners. The cost of legal preparation for these actions by engineering department staff alone could be reduced by an estimated $50,000 per year.

Future and external benefits include improvements to the environment as a result of resolving the basement flooding more quickly. Currently, when the sewers back up, their flow is diverted to the storm sewers, which are then diverted to the waste processing plant. Sewer backups substantially increase the waste processing plant's volume. The costs of processing could be reduced by $10,000 per year.

These benefits alone could result in savings of over $160,000 per year. If you expand that calculation over 10 years, the benefits would be over $1,600,000, not taking inflation into account. These figures strengthen the case for building the new databases. Gary can now compare the benefits that will result from the new information against the costs of data acquisition and system implementation.

The sample forms used in this case study have been used for the successful collection of information product descriptions in state, regional, and municipal organizations over many years by Tomlinson Associates Ltd. They can be found on the ESRI Virtual Campus course by the author called "Planning a GIS." Visit campus.esri.com. Examining these forms will help you appreciate the comprehensive nature of the information needed to ensure that requirements are fully evaluated.

At some stage in running GIS all of this information will be needed by the GIS manager to plan his/her day-to-day work. The thinking is going to have to be done sooner or later to run the system or explain the work to users and senior management. Doing this thinking when faced with daily production demands in a production GIS shop is a recipe for GIS manager madness (at least frustration and overwork). It is better to do it before production begins (during planning) and not after.

Title: Sewer backup information product
Required by: Engineering dept.
Name: Gary

Savings

Current data compilation is time-consuming—100+ person hours for each flooding (50 per year). This time could be reduced by 90 percent. Staff time savings of 80 percent per year on current workload.

Benefit to agency

Timing of product output will be improved significantly. Information provided on basement floodings just after the storm occurs. Solutions may be applied in time to verify the correction during the rainy season.

The improved timing will make it possible to coordinate repairs with maintenance contracts currently underway, which would save considerable sums.

Benefit—$100,000/year

Reduced liability. There is currently a backlog of 10 court cases awaiting trial. The number of court cases increases as time goes on. Staff spends considerable time collecting data for court cases. Legal costs would be reduced.

Benefit—$50,000/year

Future and external benefits

By resolving basement-flooding problems, the environment is improved. To relieve basement floodings, sanitary sewers are pumped into the storm sewers, and there is an increase in pollution levels in fresh water streams and lakes. Sewage treatment costs would be reduced.

Benefit—$10,000/year

Define the system scope

"Scoping the system means defining the actual data hardware, software, and timing."

Define the system scope

Defining the system scope is the stage where you answer the following questions:

- How much data is needed?
- What hardware and software is needed?
- When must the system be put into place if the organization's requirements are to be met?

In chapter 6 you learned how to create information product descriptions, called IPDs for short. IPDs are important planning tools that collate and summarize the essential information about what users want out of the system. In this chapter, we take a more detailed look at what should go into the system using a summary document called a master input data list (MIDL).

As you created the individual IPDs, you identified the data needed. Now you're going to use this information to create a new document—the master input data list. The MIDL is a detailed list of all the data sets that must be entered into the GIS system to generate all the information products needed. The MIDL should identify each data set (with its name, ID number, and source agency name) and include assessments of the data volume (amount) and format, and the availability of source data. This data about the data has a name, too. It's called metadata and today's modern GIS systems have built-in metadata handling capabilities.

The MIDL:

1. is the **master** list of data that will go into your system
2. spells out work that will be involved in putting the data into your system

The MIDL should include guidelines or estimates where available about such things as the format and volume of digital data or the amount of digitizing required, the amount of alphanumeric data input required, the amount of data that will need to be available each year.

To be included in the MIDL, each data set must be needed and specified for at least one information product. No other data should be described in the MIDL. If someone comes to you offering "good" data to include in the GIS, ask that person, "Which information product needs this data?" Even "good" data should not make it into the MIDL unless it is needed for the creation of at least one information product. Having this business rule in place forces people to think in terms of the end products from the GIS; without this rule, your GIS data directories will quickly grow into a confusing array of layers and features that nobody understands because they don't ever get used.

With experience, you'll learn to create the master input data list at the same time as you create information product descriptions. This way, as you identify the data sets necessary to create information products, you can record the name and characteristics of the data sets directly onto your MIDL as you go.

A member of the GIS planning team should be responsible for creating the MIDL. In many cases, it will be appropriate for the team leader to undertake this task. Whoever is responsible, that person should have a good knowledge of the characteristics of data available in the organization and know where to get more information about other data when necessary.

A master input data list should include the following four components:

1. Data identification details

This is the information that uniquely identifies each data set. Because different people or departments may have different names for the same data, it is important to create a standard common identification scheme.

2. Data volume (amount) considerations

The volume or amount of data you need will impact your system design. Spatial data tends to demand a lot of disk space; having adequate, scalable storage will save you grief down the road. If you're lucky, some of the data will come spatially referenced in the format and scale and projection you need it. But more realistically, there will be much data that requires conversion from another format. Some will not even be digital data; these nondigital data sets will require conversion to a GIS-readable format (a process known as data automation). The amount of data you need to store and also the amount you'll have to preprocess affects your working strategy for data storage and handling. You should evaluate the overall number of data sets you need along with their size, and the number and size of their attribute files. Will you need two data sets or two hundred? Are the attribute files associated with these data sets large or small?

3. Data characteristics

Understanding the format and pedigree of the data you intend to be incorporating into the GIS will guide your database design and ultimate selection of software and hardware. You need to know whether the data is available as digital files already in an appropriate GIS format. Is the format one that you can use directly in your system or do you need to convert the data from another format before using it? What are the data characteristics that will determine your input method (scanning, digitizing, or text input)?

4. Source data availability and cost

It is important to know the availability of data and its cost as you to develop your implementation strategy. Sometimes, it is cheaper and easier to recreate data than to edit and update existing commercial or government data. Some of the questions you need to answer now are:

- Does the data exist in a digital format or will you need to automate it?
- What is the cost of acquisition? Do royalties need to be paid?
- Is it possible to partner with another agency to share the cost of data?
- What is the cost of converting data from one format to another?
- Who is going to be responsible for updating and maintaining data?
- Are there any restrictions on the use of data?

Determining the items needed in your master input data list

The information items needed in a master input data list (MIDL) will vary with each new project and each new type of data, but the general rule of thumb regarding metadata is to capture everything that will impact your acquisition and use of the data you need. Metadata doesn't take up much storage, but taking the time to set it up correctly will pay dividends down the road when stakeholders start questioning the basis of your analyses. Any book on IT planning would be remiss not to drag out the shopworn phrase "garbage in, garbage out" to drive home the importance of stocking your GIS with only the most reliable and accurate data. This is not to say that lower resolution, or more generalized spatial data cannot be loaded

Chapter seven

Component	Details needed	Notes
1. Data identification	Data set name Data set number Source agency name Internet location (URL) Metadata available	 Yes or no
2. Data volume considerations	Source data medium Digital data format Percentage available now in digital format Primary record type Primary record volume Total data volume	 Line, sheet, etc. Number of primary records
3. Data characteristics Scanning considerations	Sheet size Minimum scan resolution Legibility	In in. or cm, typical and maximum In dpi, without compression Percentage of total data volume in high, medium, low, and difficult categories
Graphic portion	Size Schematic Photo image Map projection/datum Measurement (COGO) data volume	In in. or cm, typical and maximum Yes or no Yes or no Typical and maximum volume or coordinates Typical and maximum number and size of observations
Digitizing effort	Polygons per sheet Lines per sheet Points per sheet	Typical and maximum In in. or cm, typical and maximum Typical and maximum
Text portion	Lines per sheet Data elements per record Total alphanumerics per sheet on input	Typical and maximum Number of fields per record, typical and maximum Number of keystrokes Typical and maximum
4. Data availability and cost	Percent coverage available now Currency Restrictions on use Cost of data set acquisition Royalties	Or date when it will be available Date of data capture and date of most recent update On acquisition and use

into a GIS, but rather that if such data is to be used that the associated metadata reflect any limitations.

The items to be listed on the MIDL detail the data source, volume name, any conversion necessary, the outright cash cost, and any availability schedules. The following list summarizes the information that you may need to collect for each data set in the master input data list. You should add or delete items in this list as appropriate for your own project.

Evaluating basic system capabilities needed to input data

Once you've collected the information for the master input data list (MIDL), you can evaluate the basic system capability functions necessary to put each data set into the GIS.

For each data set in the MIDL, make a list of the functions that are necessary to get the data into the GIS. The functions for each data set will differ depending on the type and characteristics of the data to be input.

The graphic below shows the basic functions necessary to input a digital image and a hard copy map. The individual functions only need to be listed once for each data set.

	Data needed	Data-handling functions required
Data set 1	Digital image data (remote-sensing image supplied on CD-ROM)	File transfer Reformat Create and manage database
Data set 2	Hard copy map data (Property parcel boundaries available only on map sheets)	Digitize Reformat Create topology Add attributes Create and manage database

Once the list of functions required to handle the data has been created it can be combined with the list of functions identified on the IPDs (see chapter 6) to complete the list of software functionality that must be provided by the competing software vendors.

Data inventory

Interviewer: _____ Date: _____
Name: _____ Department: _____
Phone#: _____ Division/Office: _____

Data set name:_____ Data set number:_____
Source agency:_____ Metadata? (circle)   Yes   No
Data format: (circle)   Map   Manual data file   Automated data file
                           Air photo   Imagery
Other: _____ Source data medium: _____
Digital data format: _____
Percentage available now in a digital format: _____
Total data volume: (# of pages, CDs, maps, etc.): _____

Scale of source data: Map projection/datum: _____
Photo image?   Yes   No
Digitizing effort (per sheet): # of polygons: _____
                                  # of lines: _____
                                  # of points: _____

Date of data capture: _____ Date of last update: _____
Percent of coverage available now: _____
Date entire coverage is available: _____
Restrictions on use: _____
Cost of data set acquisition $:_____
Royalties on acquisition and use $: _____

Collecting information for the master input data list

Because some of the information required for the master input data list (MIDL) is detailed and requires time to assemble, it is a good idea to be already collecting information for this document while preparing the IPDs. The form above includes a place to enter information about data sets identified during the IPD-building process. Don't get hung up trying to answer all the questions right away. The idea is to give each data set a name and a number and describe the data sufficiently so that it can be recognized. Later, you can return to fill in the other portions of the form. As you can see, this form covers most everything you can think of to describe data including data set name, source, scale, projection, format, description, and volume.

It's good practice to attach sample data to the form. This might include representative maps or all the columns for two to three records of tabular data, etc. These samples will serve to clarify the functions that the user must perform to create or use the data.

Data samples should be collected together in a physical place, what we call a "data shoe-box." If data is not represented with a sample in the shoe-box it is considered nonexistent in terms of the GIS. Many people new to GIS are unclear about what data exists in their organization. The shoe-box is the test! Use a separate data inventory form for each data set.

Basic system capability input functions needed to create the sewer segment map

Continuing the example from chapter 6, Gary has been assessing the process that will be necessary for him to create a sewer segment map. This is one of the components necessary to create the sewer incident report. This segment map needs to include the location of each sewer segment together with the sewer segment numbers. Two data sets are needed it. Both data sets are listed on the master input data list. Now Gary must evaluate the basic system capability functions needed to merge the two data sets. First, Gary outlines the steps needed to input the data. Example of two of the data sets are given below:

Data set 1: Sewer plans on paper maps (11,000)
1. **Scan**—to create digital raster files
2. **Raster to vector**—converts cells to lines
3. **Edit and display**—corrects errors introduced during conversion
4. **Create topology**—produces sewer network
5. **Add attributes**—adds sewer segment numbers
6. **Edge match**—creates single seamless digital file
7. **Create and manage database**—implements a tiling storage structure for easy access
8. **Update**—maintains sewer segment data

Data set 2: Aerial photographs in digital format
1. **File transfer**—copies aerial photographs to the GIS database server
2. **Use handheld GPS**—to create a vector file of manhole locations (points)
3. **Add attributes**—adds manhole characteristics

4. **Symbolize**—adds appropriate manhole symbols for display

5. **Rubber sheet**—stretches to fit sewer to known position of manhole

6. **Create topology**—incorporates sewer segments into sewer network

7. **Update**—includes new sewers or manhole locations as added

These steps provide the list of the basic system capability functions needed for each of the data sets. You only need to make note of them as follows:

Sewer plans
Scan
Raster to vector
Edit and display (on input)
Create topology
Add attributes
Edge match
Create and manage database
Update

Aerial photographs
File transfer
Input GPS points
Add attributes
Symbolize
Rubber sheet
Create topology
Update

These functions and those from the other data sets will be added to the list of the information product functions generated during the creation of the IPD.

Setting priorities

Next we've got to prioritize because all the information products that you require from your GIS cannot be produced at once. You need to prioritize your list according to the relative importance in contributing to the organization's objectives. These priorities will be ranked by using a scoring method, a group-consensus method, or some combination of the two. External consultants should never be allowed to do the prioritizing of the information products; they don't have enough of a vested interest in the missions and objectives of the organization. Remember, this ranking is the thing that will dictate which information products get delivered first, so make sure the process you use incorporates the appropriate input of upper management.

It is essential that during prioritization you rank information products in strict numerical order and that no two information products have the same rank. There are two methods for assigning priorities:

1. The scoring method

The GIS team leader creates a simple model that allows scores to be given to each information product based on the benefit it will provide. Scores can be assigned based on ease of production, relevance to the organization's strategic plan, or any other criteria that the team decides are appropriate. The GIS team leader can allocate scores initially alone or in consultation with other members of the planning team.

The list of scores that results is presented to the GIS planning team and to senior managers and decision makers for comment and approval. At the end of the day, senior management should make the final decision on priorities.

2. The group-consensus method

This is a less structured approach to assigning priorities. When using this method, you need to get all the managers and decision makers into a room and work together until a consensus is reached on the priorities assigned to all the information products.

Consultants should never participate in this stage of the GIS planning process. The priorities placed on the information products are the concern of the organization. As conditions change, priorities may need to be adjusted in the future, but this initial assignment of priorities gives essential direction to your GIS implementation.

Determining system scope

In this section, you will also begin to evaluate the scope of your system, which involves considering:
- data-handling load, workstation requirements, and location
- data storage and security requirements
- data readiness for use
- data priorities

Data-handling load

Once you know what the priorities are, you can use the IPD and MIDL to get a quantifiable estimate of the data-handling load that your system needs to support. The data-handling load is simply a factor of the number of products that the system will create per year and the amount of work that the system has to do to create each product. The data-handling load provides a guideline that can be used to determine the computational power and storage dimensions of the system that you need to implement.

The amount of computational work required to produce an information product depends on processing complexity and data volume. At this point in estimating data-handling load, it is only necessary to make rough estimates of processing complexity and data volume.

Processing complexity is either high or low. It can be estimated from the number of steps needed to make the information product, weighted (mentally) by the number of advanced functions (e.g., topological overlay, network analysis, 3-D analysis) that are used. The functions which are considered to be complex are so indicated in the lexicon in the back of this book.

Data volume is either high or low. It is the amount of data that the system would use in producing one information product. It can be estimated from the number of different data sets needed to create the product weighted (mentally) by the volume of data (number of items and size of area) that must be extracted from each data set to make the information product.

It is only necessary to assign a high or low evaluation. For some information products, this will be an obvious selection, for others you'll have to use your judgment. An estimate is adequate at this stage in the planning process.

You should go through your list of information products and assign a high or low value of processing complexity and data volume to each one. You can now estimate the type of workstation that will be required for each information product. Three workstation types should be considered:

1. Multi-user Windows® Terminal Server Workstation (approximate cost in 2003 = $12,000 + $8,000 software—for use by up to 15 persons)
2. High-end Windows 2000 Pentium® 4; 2.6 GHz; 21-inch screen (approximate cost in 2003 = $4,500)
3. Standard Pentium 4; 1.5 GHz; 17-inch screen (approximate cost in 2003 = $2,000)

The following table indicates which workstation type is suitable for different combinations of processing complexity and data volume.

Based on the data processing complexity and data volume used, and using your own experience or with the advice of a consultant, you can now estimate the amount of workstation time needed to make each information product. From the number of times each information product is needed per year, calculate the total number of workstation hours necessary. You can use this to make a preliminary appraisal of how many workstations of each type are needed in each department.

Occasionally one information product will use all the available hours on one workstation. More frequently several information products may be produced from a single workstation or be made at different workstations in different parts of the department in different locations.

Terminal servers (also known as client-server systems) are usually employed where the data-handling load is high and multiple users need access to the same data. They are particularly appropriate for very large data sets.

Processing complexity	Data volume to be handled	Type of workstation required
		1. Multi-user Windows Terminal Server Workstation 2. High-end Windows XP Pentium 4 3. Standard Pentium 4
High	High	1 or 2
High	Low	2 or 3
Low	High	3
Low	Low	3

Data hosting and user locations

The locations where (on what machines) data will be hosted and where users will operate from both impact network communication requirements and should be thought out in advance. For the purposes of estimation, it is helpful to assume that an information product will be generated in the user's department, although in the era of Web services, this is often not the case.

If you are a single user in a single department with a single computer using either high- or low-complexity processing and not hosting a Web site, this part of the analysis is extremely simple. On the other hand, if you need to consider the location of numerous databases and users with different types of workstations and in different departments, the situation is more complex. We examine this more in the Distributed GIS and Web services section in chapter 10. For each department, whether in the headquarters building or at a remote site, you need to calculate:

- the total number of workstations using GIS, first for high-complexity and secondly for low-complexity processing (if both, use high)
- the number of users who will be using their workstations simultaneously (concurrently) at the time of peak usage

If any of the departments are hosting a Web page based on GIS data, you also need to estimate the number of visits per hour (or hits per hour) that are anticipated. (Note: either measure is helpful. Hits are best but more difficult to estimate. You should use one or both in your analysis.)

Here is a table showing the use estimates for three departments in headquarters and two remote offices. It shows that among all locations and departments, as many as 40 people will be considered high-use clients, with as many as 14 needing simultaneous high-use access. There are 82 people who need low-use access, with as many as 29 needing it concurrently during peak hours.

It is also useful at this stage to make some notes on the computer system with which you, the GIS leader, will access the system. The GIS leader usually has special super-user privileges and specialized mission-critical, rapid-response applications and should be equipped with a high-performance machine.

Location		Processing complexity				Intranet/ Internet
		High		Low		Visits or hits per hour
		Total users	Peak concurrent users	Total users	Peak concurrent users	
Headquarters	Planning	8	2	16	5	250
	Engineering	12	5	5	3	-
	Operations	2	2	30	10	-
	Totals	22	9	51	18	250
Remote sites	Exning	10	3	21	6	500
	Gazeley	8	2	10	5	400
	Totals	18	5	31	11	900
	Grand totals	40	14	82	29	1150

Data storage

The type and cost of disk space that you require depends on the volume of data that your system will handle and your security requirements. You should be keeping track of data volume amounts within the MIDL. Use the total data volume recorded in the MIDL to estimate the actual amount of disk space required by the information product.

Data volume is calculated as the number of gigabytes of storage space required and can be classified as follows:

Data volume	Storage space required
High	Over 1,000 gigabytes
Medium	40 to 1,000 gigabytes
Low	Less than 40 gigabytes

Yours may be an organization that requires data security measures of some level to protect against accidental or deliberate loss or corruption of data. Most organizations also restrict access to sensitive data to prevent misuse. A network-based mass storage system can be secured against loss, for example, by including mirror sites as backups as well as tiered username and password protection. Stand-alone personal computers offer much more limited security features. Computer security levels can be classified from high to low:

1. High security: Full mirror, RAID level one (highest) security (100 percent redundancy of data on mirror; no need to rebuild data in the case of disk failure).
2. Medium security: RAID level five security (a much lower level of security than RAID level one). A disk failure will have an impact, and it is difficult to rebuild a database in the event of a failure.
3. Low security: Tape or compressed disk backup (you're only as good as your last regular backup.)

The appropriate amount of storage and security breaks down as follows:

Chapter seven

Data storage options	Disk space	Security	Approximate cost (2003 prices)
Enterprise network mass storage	600 gigabytes to 10 terabytes	High security	$35–$80 per gigabyte
Work group server	40–1000 gigabytes	Medium security	$19 per gigabyte
Personal computer	40 gigabytes	Low security	$3.75 per gigabyte (IDE) $17 per gigabyte for SCSI, 1500 rpm

Data priorities

In the same way that you ranked the information products by priority, you should rank the data sets based what data the master input data list is telling you need first.

You should reorder the master input data list based on the order in which data is required to make the information products. On the new version of the list, data sets that will be used first will have the highest priority. The end product of this simple step should be the assignment of data priorities to each data set in the master input data list. The list itself should be reordered with the highest priority (most immediately needed) data on top.

Data readiness

There is an important difference between data availability and data readiness. Data availability is the date on which you can receive data from its source. This date follows the acquisition or gathering of data (including negotiation of any agreements about data use). Data readiness is the date on which the data is in your system, processed, and ready for use in the creation of an information product. All data entry, editing, reformatting, and conversion processes are complete.

Scoping hardware requirements

At this stage of the process, you need to take a quick look at the basic hardware requirements so you can begin to think intelligently later when asked to do so. In the context of this discussion we'll use as an example

ESRI's leading line of GIS software. The basic options for computer systems can be thought of as a three-tier option.

- Servers: multi-user UNIX® or Windows server/workstations
- GIS workstations: high-end Pentium 4 workstation with lots of RAM for running core GIS software (i.e., ArcInfo™, ArcGIS, ArcServer)
- Standard Pentium 4 workstations: for Web access to GIS resources and (thin) desktop-client applications

System performance expectations have changed significantly over the past few years. In particular, better platform performance and lower hardware costs have affected productivity.

What affects timing?

Careful planning of time and resources is essential to effectively manage the complex process of creating information products from scratch in a logical sequence. To plan for successful GIS you need to understand the factors that can affect the readiness of data when it is needed and how these factors impact the scheduled production of information products.

In this section, the following factors that affect project timing will be discussed:

- Data input timing
- Information product application programming timing
- Product demand timing
- System acquisition timing
- Training and staff issues

How you get your data into the system will have a major impact on your timetable. If you've got to create data like most new GIS efforts of any size, the data that you will be creating yourself is likely to require significant human labor. Each method takes a different amount of time, and each is suited for a certain type of data. The four most common methods of data entry are digitizing, scanning, keyboard input, and file transfer.

Recall the sewer segment map from the previous chapter that was needed by the engineering department to link the sewer segment numbers, characteristics, and street addresses in the database. The high estimated cost of producing this new database was incurred in manual data entry: the scanning of thousands of old sewer plans/profiles; the digitizing of real positions of the manhole covers based on GPS-gathered data, the registration of images with the digital points, all leading to ultimate creation of a sewer network map. The cost of the production of this data is largely attributed directly to labor.

In a typical municipal GIS, data entry can consume up to 80 percent of the time it takes to implement a GIS project, so if large amounts of data must be created, make sure you allow ample time in the schedule for that to take place, because GIS is a rocket that won't launch without fuel in the tank.

Before you start creating your own data, you should check to see if you can obtain the same data more quickly, more affordably, and to appropriate quality standards from elsewhere. Common sources of data include:

- similar or partner organizations
- local and central governments
- census and survey organizations
- commercial data providers

If you obtain digital data from other sources, reformatting and editing are frequently necessary to allow the data to be used with existing data sets. Most GIS software can work directly with a variety of data formats. The following table shows the names and acronyms of some commonly used vector and raster data formats.

If your software does not directly support a certain data format, you may be able to use a data conversion utility to allow its use in your system. Many software packages include a variety of data conversion utilities. If your package doesn't include the conversion utility you need, you may be able to purchase it from a data supply company.

Based on your own experience or with the advice of a consultant, you should estimate the number of weeks that it will take to move from data availability to data readiness for each data set. Given the date of availability, determine the first calendar date at which data could be ready. The data-readiness date will probably be revised based on need during the activity planning that follows. However, it is useful to have a potential data-readiness date in your mind as you start activity planning.

The development of custom applications with a large number of steps is sometimes required to make an information product. Customized application programming is necessary to ensure that products can be created quickly and efficiently. Other custom programs may also be needed if there are time demands that put time limits on product creation.

Vector formats	
ArcInfo coverages	Interactive Graphic Design Software (IGDS)
ArcInfo export files (E00)	
ArcInfo geodatabases (introduced in ArcInfo 8)	Initial Graphics Exchange Standard (IGES)
ArcView® GIS shapefiles	Land-use and Land-cover data (GIRAS)
Atlas GIS™ Geo .agf files	
AutoCAD® drawing files (DWG)	Map Information Assembly Display (MIADS)
AutoCAD drawing interchange file (DXF)	MicroStation Design Files (DGN)
Automated Digitizing System (ADS)	MOSS export file
Digital Feature Analysis Data (DFAD)	S-57
Dual Independent Map Encoding (DIME)	Spatial Data Transfer Standard (SDTS)
	Standard Linear Format (SLF)
Digital Line Graph (DLG)	TIGER/Line® extract files
Etak® MapBase® file	Vector Product Format (VPF)
Raster formats	
Arc Digitized Raster Graphics (ADRG)	GRASS (Geographical Resource Analysis Support System)
ArcInfo GRID	Grid
BIL, BIP, and BSQ	IMAGINE®
BMP	JFIF
DTED (Digital Terrain Elevation Data)	JPG
	RLC (run-length compressed)
ERDAS®	SID files
GIF	SunRaster™ files
	Tag Image File Format (TIFF)

The IPD is the starting point for application development. The "steps to make the product" captured in that document are used to specify the application. You should test the resulting application program thoroughly in a production environment before you consider it ready for deployment in a front-line situation. The development of application programs might be done by you or your staff, or outsourced to contractors. If outside contractors are used, you will need to allow additional time for the contracting process.

Based on your own experience, or with the advice of a consultant, you should estimate the application programming time needed for each

Chapter seven

Size of application program	Nature of application program	Time needed for development
Small	No application program needed Modest number of steps required to create information product	Zero One staff month
Medium	Straightforward information product required, but a large number of steps	Up to 12 staff months or 3 months elapsed time
Major	Application involving many steps in a time-sensitive situation (e.g., a license permitting process)	Up to 144 staff months or 12 months elapsed time

information product. This will vary depending on the complexity of the application required:

If writing and application is projected to take 12 months elapsed time, it is wise to break the application into two or more stand-alone pieces. The rate of failure of over 12-month projects is high.

Your application programming time estimates at this stage of planning should be considered to be plus and minus 50 percent. A programming activity that you estimate will take two months may actually take between one month and three months. You should also bear in mind that longer application programs (over 12 months) have a high risk of project failure as a result of changing technology lifecycles.

Other factors influence the timing of the generation of a product. These include the rate at which products can be generated, the rate at which products can be used by the organization, and the potential for changes in information product priorities because of multiple data use.

Manage expectations from the beginning. There can be delays as you get up and running. A specific plan for information-product delivery will be addressed when you start activity planning at the end of this chapter. You should think about the rate at which the information products generated by the system can be absorbed and used within your organization. If you are thinking of producing two hundred new maps per year (of anything), do existing staff have time to examine and use two hundred maps? You need to go over the high-priority information products and discuss their use with staff involved to ensure that the rate of use matches the proposed rate of generation. The results may be surprising, especially given the work that has gone into information product description. If the demand picture changes, adjust your production rates accordingly.

Sometimes, in the process of inputting data to create high-priority information products, sufficient data is entered to generate a lower-priority information product. This is a happy by-product and should alert the GIS planning team to try and identify other potentially multiproduct leveraging. Sometime it may even be beneficial to alter the information product priorities to maximize multiple data-use potential.

Time must also be allocated for system selection and system procurement.

Government procurement procedures may be among the slowest. They can involve any or all of these steps:

- Issuing requests for information (RFIs) and requests for proposal (RFPs)
- Proposal review
- Benchmark testing
- Negotiation of best and final offers
- Timing of budget cycles

Each of these procedures may have its own requirements for delay and acceptance. The amount of time involved depends on the organization involved. You should estimate the time needed based on your own experience and after consultation with your budgeting and procurement manager in your organization.

Finally, allow time for system implementation and a reasonable break-in period, regardless of the targeted operating efficiency of your GIS. During this time staff must work through a steep learning curve and inevitable problems and glitches have to be resolved. It may not be possible to reach full production status immediately, so there is an impact to the immediate rate of production. Learn to manage expectations up front. System break-in will also be facilitated if you can allow a window of relatively low demand at the beginning.

Time for training and staffing issues should be included in your implementation plan. Training needs vary depending on the types of GIS users that need to be trained.

Within large GIS-using organizations, you will find the following types of users:

- Professional GIS users supporting GIS project studies, GIS spatial data maintenance, and commercial map production operations.
- Desktop GIS specialists supporting general spatial query and analysis studies, simple map production, and general-purpose query and analysis operations.
- Business users who require customized GIS information products to support their specific business needs. These are end users who do not require geographic expertise and use information products to support standard business functions.
- Internet and intranet map server users of simple map products using simple publishing wizards and intranet and Internet browser clients.

All four groups of users require training. In addition, new staff may need to be recruited and brought onto the GIS team. Training is a vital part of GIS and a significant budget item. It is an essential part of maintaining and keeping skilled GIS staff. Provisions for this must be included in your time plan.

Activity planning

If you've followed the procedures outlined in this chapter, you should now have the following:

- Information product priorities (perhaps modified somewhat by multiple data-use potential)
- Data priorities
- Data input timing
- Application programming timing
- System acquisition timing
- Training and staffing timing

Now you are ready to start the overall activity planning for your GIS. The first step is to revisit the information product priorities to answer some questions:

- Will some information products depend on the creation of other information products? If so, the latter must be assigned a higher priority.
- When will it be possible for you to make each information product? When, given your estimated data readiness dates (which include both data availability and the time required to input the data)? This is particularly important for the high-priority information products that you want to make early.

Use these considerations to revise your information product priorities and establish a second-level priority review.

The next step in activity planning is to use a Gantt chart—essentially an illustrated timeline that includes the ability to plot dependencies on a linear chart so that one can see which activities are time-senstive. Gantt charts can help you plan and manage your project effectively. There are many other project evaluation and review techniques and specialist project-management software packages, but this is one of the simplest and easiest to apply. You need to manage your project so that you are focusing on doing the tasks that are going to produce results.

A Gantt chart displays information about activities and their duration. It allows you to schedule and track the activities associated with GIS planning and implementation. Each data set or information product, system acquisition, or staff activity is a row in the chart, and time periods are columns in the chart. Bars are used to indicate the time necessary to complete each activity, e.g., data-loading application, writing break-in period or staff holiday. Gantt charts make it easy to see when a particular activity will be completed and when other activities relying on this completion can begin. For example, the entry of a particular data set might be one activity, and at the end of this activity it would be possible to produce a particular information product.

You can use a Gantt chart to document project activities and their duration, establish relationships between your activities, see how changes in the duration of activities affect other activities and track the progress of your project. You will also use the Gantt to schedule hardware, software, and network acquisitions over time when these have been determined.

Gantt charts can be constructed using ordinary spreadsheet software such as Microsoft Office Excel and can also be found in specialist project-management software from a variety of suppliers. They can also be made by hand, where the situation is relatively simple.

The list of activities on the left of your Gantt chart for GIS planning should include the name of every data set that you intend to put into your system in the time period covered by the chart. The list of data sets should be followed by a list of names of the information products that you intend to generate in the same time. Rows for system acquisition, system start-up, system staff training, and holidays and sick leave should also be included.

The columns on the top of the chart are usually best divided into one-week segments grouped into months and years. You should indicate fiscal years or budget cycles. The chart usually covers a five-year period.

In general, the order in which you complete your Gantt chart is:

1. system acquisition and staff training activities
2. data-input activities—establish when data will be ready
3. application development activities for any high-priority information product

Gradually you build your Gantt chart, taking into account restrictions of staffing and product demand, until you have decided how and when to build each information product.

Now you have, for the first time, a date on which you can reasonably anticipate the delivery of the first of each information product. In chapter 11, through benefit-cost analysis, you'll learn that you should allow one year from the date of the first of each information product to count the benefits that accrue from having that product available.

Create a data design

"*The data landscape has changed dramatically with the advent of the Internet and the proliferation of commercial data sets. Developing a systematic procedure for safely navigating this landscape is critical.*"

Create a data design

Data characteristics

Part of developing a systematic procedure for creating the conceptual system design is having a thorough understanding of the characteristics of your data. These characteristics include each data set's scale, resolution, map projection, and error tolerance, and how the data affects the intended information products. Sometimes you'll need to create multiple versions of the same data sets at different scales or resolutions to create the map output specified by the IPD. These will all be stored as individual layers in your database.

Scale

Scale is the relationship between the distance on a map and the corresponding distance in the real world. If a map has a scale of 1:24,000, then one inch on the map is equal to 24,000 inches (2,000 feet) on the ground. Map scale can also be expressed as a statement of equivalence using different units; for example, one inch = 2,000 feet. The scale of data reflects resolution and its relative accuracy on a map: the larger the scale, the more accurate and detailed the data set. Map scale numbers are counter intuitive: the higher the number the lower resolution. So 1:6,000 data is significantly higher resolution than 1:100,000 data.

Features drawn in large scale show greater map detail because they have more room on the paper and contain lots of information. On a small-scale map the features are usually generalized or aggregated. Both are useful and serve a purpose. Consider the following illustration. A small-scale data set would show a river as a single gently curving line while a large-scale data set would show a smaller part of the small river as a polygon depicting the banks and the width of the river.

What scale or scales to set up in your GIS database (i.e., how high a resolution of data to acquire) is a critical factor in the system design. If the scale of primary data is too large, the data volume in bytes could overwhelm the computational processing resources. This becomes especially significant when you're trying to build quick-response systems. If the scale is too small, the database may not be able to perform the best analysis or

Chapter eight

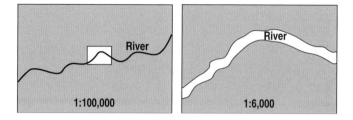

reliably provide the specified information products. The "native" format of your source data as well as the budget resources at your disposal to purchase computing power will advise this process.

Once you've chosen the appropriate scale for your GIS database, you may need to convert some data to the common scale either in advance or in real-time as one of the steps required to create information products to fit with other data.

A good rule of thumb if changing scale is to avoid changing scale by more than two and a half times in either direction. For example, if you start with a 1:50,000 scale data set and reduce the scale two and a half times smaller (multiply 50,000 times 2.5), your scale will be 1:125,000. Thus, if you create information products from that source data that have scales smaller than 1:125,000, you may run into data profusion and legibility problems. If you start at 1:50,000 and go two and a half times larger (divide 50,000 by 2.5), your maximum scale will be 1:20,000.

In some cases, you may need to store data at more than one scale in the database. Storing source data at more than one scale is required when applications will be performed at both small and large scales. Sometimes you may want to store base-map data acquired from external sources at the native scale at which it was received. Although a GIS does not typically store the scale of the source data as an attribute of a data set, scale is an important indicator of accuracy, which is why you note it in the IPD. All users of the data must understand that true spatial analysis—like when you're using geographic overlay functions to derive new information—is only as accurate as its highest resolution layer. If more than one scale is represented in the database, it should be well-documented in the meta-data. Note that if the different scales are from different sources, they may not agree.

Scale influences both the cost and accuracy of the resulting database. For example, the number of map sheets needed to cover the same area in your database increases exponentially with the scale. Therefore, mapping at 1:6,000 is 16 times more expensive than mapping at 1:24,000. As the number of sheets needed for your application increases, so does the cost. And because scale does introduce such a significant impact on cost, you must take a close look at the actual needs as specified in the IPDs. Don't build a rocket to Mars if all you need is one to the moon. The purpose of the application will determine the appropriate scale.

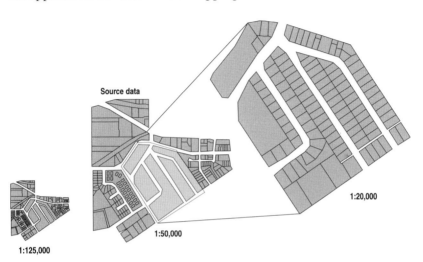

Resolution

Resolution is defined as the size of the smallest features that can be mapped or sampled at a given scale. The resolution of a map is directly related to its scale. As map scale decreases, resolution diminishes and feature boundaries must be smoothed, simplified, or not shown at all. There will be a minimum polygon size and line length you can represent at a given scale. Features below these resolutions are merged into surrounding data, converted to a point, or deleted. In the following graphic you can see that a 10-acre polygon is visible at a scale of 1:24,000, but the same polygon appears as a point at 1:500,000. Resolution also determines the distance between sample points in a grid or lattice format (for example, satellite imagery).

Like scale, your data must have the minimum resolution in order to create your information products. For example, a city's land-parcel data, must be high resolution, while a Web application showing interstate travel routes would have to be small-scale. But keep in mind that resolution also

Chapter eight

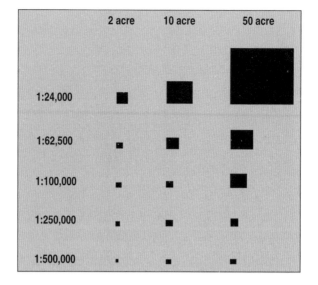

contributes to data error. Your information product descriptions include error tolerance and will help you determine the required resolution. You do not necessarily need the higher resolution available.

Map projection

The choice of a map projection is another crucial step in the GIS database design process. As part of the conceptual system design, you need a basic understanding of map projections in order to determine the best map projection for your needs.

Paper maps are perhaps the most common source of geographical data in the world. Because the planet Earth is spherical and maps are flat, getting information from the curved surface to the flat one requires a special mathematical formula called a map projection. A map projection simply converts Earth's three-dimensional surface to the flatlands of paper—a decidedly two-dimensional space. The process of flattening the earth creates map distortions in either distance, area, shape, or direction. The result is that all flat maps have some degree of spatial distortion. The type of projection used determines the degree and type of distortion found on the map. A specific map projection can preserve one property at the expense of the others, or it can compromise several properties with reduced accuracy. You should pick the distortion that is least detrimental to your database when selecting a projection type. Today's GIS programs offer an array of projection choices, but read the supporting documentation and help files before making any database decisions or attempting to reproject any data.

Datums are another important map aspect related to projection. A datum provides a base reference for measuring locations on Earth's surface. It defines the origin and orientation of latitude and longitude lines and assumes a shape for the globe on which they fit. The most recently developed datum is the World Geodetic System of 1984 (WGS84). It serves as the framework for satellite-based location finding worldwide (i.e., the Global Positioning System—GPS). There are also local datums which provide a frame of reference for measuring locations in a particular area. The North American Datum of 1927 (NAD27), the North American Datum of 1983 (NAD83), and the European Datum of 1950 (ED50) are local datums in wide use. As the names suggest NAD27 and NAD83 are designed for best fit in North America, while ED50 is intended for Europe. A local datum is not suited for use outside the area for which it was designed. Maps of the same area but using different datums (e.g. NAD 27 and NAD 83) will result in the same features showing up in different locations on the map—clearly an unacceptable result.

To effectively use spatial data from a map, you need to know the projection of that map and the datum. Often, input maps are in different projections, requiring a conversion or coordinate transformation. Your GIS software should support changing your data's projection and datum.

The amount of data distortion you'll experience due to map projection is related to scale. The larger the geographic area covered by the map (the smaller the scale), the more distortion from projection you will experience. The fact that a map projection results in the actual scale varying at different points on the map can be used as a guideline of distortion.

Error tolerance

Understanding the types of error you may encounter in developing and using your GIS is very important and frequently overlooked in the GIS planning process. Error, because it is related to resolution and scale, is also directly linked to cost: reducing error costs money. Everyone involved in designing and using the GIS must have a sound understanding of what error means, and what amount is acceptable and what amount of error is not acceptable. This threshold will cost a certain amount. Some error is tolerable as long as the usefulness of the information product is maintained. When people in your organization request an information product, they rightfully expect to receive benefit from using it. They either expect it to save staff time, increase the organization's effectiveness, or contribute new value to the enterprise. The product's benefit is lost if there is a lot of error in the product and it isn't useful, or worse still it becomes cost if someone makes a wrong decision based on error-ridden data.

As was introduced in chapter 5, there are four types of error: referential, topological, relative, and absolute. To review, referential error refers to error in label identification or reference. For example, in the case of a sewer-management application, are the correct street addresses on the correct houses? Are the correct sewer segment numbers on the correct sewer segments? Topological error occurs when there is a break in a needed linkage, for example where polygons are not closed, street networks are not connected. Relative error is error in the position of two objects relative to one another. For example, in a sewer-management application, imagine that a property is 60 feet wide and the roadway is 30 feet wide. The maintenance crew needs to know where to excavate. It is important to know, relative to the property and the roadway, where the sewer is located.

Absolute error is error in identifying the true position of something in the world. Absolute error becomes an issue when you are bringing different map types together in a graphical or topological overlay or combining GPS reading and maps.

Error can affect many different characteristics of the information stored in a database. Map resolution, and hence the positional accuracy of something, can apply to both horizontal and vertical positions. The positional accuracy is a function of the scale at which a map was created. Typically, maps are accurate to roughly one line width or 0.5 millimeters. For example, a perfectly drawn 1:3,000 map could only be positionally accurate down to 1.5 meters, while a perfectly drawn 1:100,000 map could be positionally accurate down to only 50 meters.

The following error table given illustrates, in tabular form, the error you can expect to have in your database given the source data, map scale, resolution, and error tolerance. Specifically, two types of error tables are utilized:

- Map scale by area and error tolerance—shows the scale at which the map must be created given the minimum area that must be measured and the percentage error that is tolerable.
- Percent error in area measurement for a given area and map scale—shows the percentage error in area measurement that can be expected as a result of the minimum area that is being mapped and the map scale.

These tables are used to show the resulting scale or percentage error if values for area and the other item are given (see table below). They assume 0.5 mm positional accuracy of mapping and an average case distribution of error.

Given information	Resulting information
Minimum area, percentage error in area	Map scale
Minimum area, map scale	Percentage error in area

To determine a suitable map scale, consider the minimum area to be measured and the amount of tolerable error. The expected percentage error in area measurements is derived from the minimum area mapped and the map's scale.

In the following error tables, you can see how map scale, the minimum size area you need to map, and error tolerance are related to one another. As you are developing your conceptual system design for data, keep these relationships in mind. For example, if you decide to map your entire database at 1:24,000 (because that is what you have readily available and 20 of your 30 information product requests require this level of accuracy), you will end up creating information products with little or no value. This will result in unmet user expectations and perhaps discredit the entire GIS. It is important to understand error and be able to effectively communicate its consequences to the people in your organization.

The concept of error tables is confusing for many people, so let's go over it again using a simple example. Marcella works at a city's economic development department attempting to lure businesses in the warehousing and distribution industry to locate in her city. These businesses typically require good rail and highway access plus relatively level land parcels of 25 acres or more. Marcella needs a map that identifies these types of sites with plus or minus 5 percent error in area. If she provides inaccurate information to new businesses, she won't get very far. As part of the conceptual system design, you will need to review her information product description, which describes her request, and determine the minimum scale for mapping parcels in order to fulfill Marcella's request. You can do this using error tables.

Map scale for a given area and error tolerance					
Minimum area (ha)	% error in area measurement				
	1	3	5	8	10
0.01	1:100	300	500	800	1,000
0.1	300	900	1,500	2,400	3,000
1	1,000	3,000	5,000	8,000	10,000
10	3,000	9,000	15,000	24,000	30,000
100	10,000	30,000	50,000	80,000	100,000
1,000	30,000	90,000	150,000	240,000	300,000
1 hectare (ha) = 10,000 m2 = 2.471 acres					

Percent error in area measurement for a given area and map scale					
Minimum area (ha)	Map scale				
	1:1,000	1:5,000	1:10,000	1:50,000	1:100,000
0.01	10.0	50.0	INVALID		
0.1	3.3	16.6	33.3		
1	1.0	5.0	10.0	50.0	
10		1.6	3.3	16.6	33.3
100	INSIGNIFICANT		1.0	5.0	10.0
1,000				1.6	3.3
1 hectare (ha) = 10,000 m2 = 2.471 acres					

First, Marcella must convert her minimum site size from acres to hectares. There are 2.471 acres in a hectare, so 25 acres is approximately 10 hectares. To determine the appropriate scale, she will use the map scale by area and error tolerance table.

She finds her minimum area in the first column of the table. Next, she finds her error tolerance (5 percent) along the top row of the table. The intersection of 10 hectares and 5 percent error is 15,000. The minimum map scale needed to create Marcella information product is therefore 1: 15,000. By using mapping at a scale of 1:15,000 or larger, Marcella's application will deliver the required positional accuracy.

Map scale for a given area and error tolerance					
Minimum area (ha)	% error in area measurement				
	1	3	5	8	10
0.01	1:100	300	500	800	1,000
0.1	300	900	1,500	2,400	3,000
1	1,000	3,000	5,000	8,000	10,000
10	3,000	9,000	15,000	24,000	30,000
100	10,000	30,000	50,000	80,000	100,000
1,000	30,000	90,000	150,000	240,000	300,000
1 hectare (ha) = 10,000 m2 = 2.471 acres					

Data standards and conversion

In this section, you'll learn the importance of reviewing your existing data, identifying data sources, developing data standards, and determining data-conversion requirements associated with your data.

Digital data sources

In the early days of GIS until even just 10 to 15 years ago, most GIS databases were created with data converted from paper form. The process of digitizing paper maps is slow and laborious. What's happened in the meantime is that more of what exists in the physical world has been measured and much of that information is now finding its way—in digital form—into the marketplace of information. Some of this information may be available from commercial vendors for a fee while other information may be available for zero or nominal fees from public agencies. Access to this world is wonderful (in fact it can really be a cost-saver if you compare what it costs to create some data), but it is imperative to have some true knowledge of the data's accuracy and pedigree (it's reliability).

On the Internet you will find a staggering array of spatial, tabular, and image data. In some countries, much data is available at little or no cost if you know where to search. Data portals, like www.census.gov, or ESRI's Geography NetworkSM, are emerging as good places to begin searching for reliable data. From private data vendors, prepackaged processed data is available for a wide range of applications. Maybe your own organization is a source of digital data. Determining what data sets meet your requirements and sifting through them takes time. The Internet and World Wide Web are invaluable tools for locating and acquiring digital data. Some of the best data is actually free, particularly from international, national, regional, and local governments. Sometimes knowing where to look and how to search is a challenge.

When determining whether or not to acquire digital data, you need to be aware of the history and quality of the data set. If you do not know the data's content, source, age, resolution, and scale, consider it useless for your purposes. You should expect to receive some metadata in the form of a data dictionary or data quality report from the provider or vendor. The metadata should provide pertinent background information about the data. Digital data can speed the process of developing your GIS, but only if you first understand what you're buying or downloading.

Technology standards

Technology standards are one of the important requirements of a successful GIS because they facilitate the effective sharing of application programs and data between offices, agencies, and the public. There are several standards to consider: operating system standards, user interface standards, networking standards, database query standards, graphic and mapping standards, and data standards. Standards regarding data, including digital data exchange formats, are the focus of this concept.

Standards bring order to the seemingly chaotic development process, and should be agreed upon early in the GIS project. Standards allow applications to be portable, and to be more accessible across networks. Along with the benefits, however, there are also costs: the time and money needed to develop and implement standards; the training of the standards, the retrofitting of existing applications; the compromise in acceptable data quality and error tolerance. For example, if you establish a standard for positional accuracy that is plus or minus 40 feet and is adequate for 95 percent of your applications, the usefulness for the other 5 percent of your applications will be compromised.

As a conscientious manager, it is your duty to consider the true benefits and costs of your program, and to communicate these costs and benefits to upper management so that funds may be forthcoming. Most smart managers realize that implementing standards costs money up front.

Data standards are developed by taking into consideration the requirements associated with your information products, the current standards (if any), and any anticipated standards that may be in effect in related departments or organizations. The current, or established, data standards in your organization are often developed through informal arrangements, continuation of past practices, and the need to deliver information products. Often, they are not documented and are inconsistently applied. Also, existing standards are often out-of-date and may not take advantage of the most recent technology.

Determining established and anticipated data standards is part of the conceptual design process. Much of the information you need is in the information product descriptions. The GIS team, representing all participants, needs to arrive at a consensus on the following standards related to data:

- Data quality standards (i.e., the appropriate map scale, resolution, and projection for source material)
- Error standards (referential, topological, relative, and absolute)
- Naming standards (layers, attributes)
- Documentation standards (minimum amount of metadata required for each data set)
- Digital interchange standards (e.g., DXF, DLG)

After the standards are developed and agreed upon by the GIS team, some organizations find it useful to formally adopt and publish the standards.

In the United States, there are national data standards being developed by the OCGIS as part of the National Spatial Data Infrastructure. It is wise to take these into consideration as you develop your system.

Survey capabilities

The GIS database can now accept measurements from survey instruments of all kinds in three dimensions, and can carry out all the traditional survey computations necessary to adjust those measurements and create coordinate points with known levels of error. Least-squares adjustment fit of disparate data can be carried out to get the best value for a point. Coordinate geometry (COGO) measurements can be included with the survey measurements and treated in the same manner. Coordinates derived from GPS stations can also be added to the survey measurement database. These survey measurements and computations are carried out within the geodatabase in the same coordinate space as other vector and raster data.

The surveying data flow from field to fabric is greatly improved. The process of moving from fieldwork station measurements through data processing of survey computations to COGO and CAD systems for drafting and design into GIS systems for integration with other data can now be handled within one geodatabase in one coordinate space and in one optimized process.

A significant new capability in terms of GIS database accuracy is to integrate the survey measurements with the location of GIS features on the map. A link can be established between coordinates established from survey measurements and points on features. Thereafter the features can be moved to their correct positions and so stored in the database. Snapping tolerances, configuration algorithms, and batch processing of adjustments can be selected. Whole new GIS features defined by survey measurements can be added, as well as improving the accuracy of the existing GIS data. Measures of error of the new feature locations can be provided by display of error ellipses. The tolerance to relative error and absolute error expressed in the information product descriptions can now be quantitatively compared to the accuracy of feature placement.

Topology

Topology is an arcane branch of algebra concerned with connectivity. It has been used since the earliest days of GIS to identify errors in the vector fabric of the database. Specifically, it could identify when polygons were not closed or when lines overshot the junction, when there were breaks in networks, when names had been incorrectly linked to features and two names had been assigned to one feature or no name had been assigned to a feature which ought to have a name. The creation of a topologically correct data set was a significant step forward in establishing the accuracy of a GIS database. It was essential, before the days of inexpensive computer monitors (cathode ray tubes), when digitizers were working blind and errors were frequent.

Topology is back with us again, after an absence of a few years, for the same reason. It is an excellent tool for establishing spatial integrity and for error identification and editing. The big difference in the current versions of topology is that they operate in three dimensions. They are multilayered topologies where features or parts of features in one layer that are coincident or intersect with features or parts of features from another layer can be intelligently topologically linked. Working with a geodatabase, rules are established to control the allowable spatial relationships of features within a feature class, in different feature classes, or between subtypes. For example, lot lines must not have dangles, buildings must be covered by owner parcels, etc. A topology is itself a type of data set in a geodatabase that manages a set of rules and other properties associated with a collection of simple feature classes. The feature classes that participate in a topology are kept in a feature data set so that all feature classes have the same spatial reference. A topology has an associated cluster tolerance which can be specified by the data modeler to fit the precision of the data.

It can apply to a limited area. Creating a topological rule does not ensure that it will not be broken, but does ensure that the error will be identified. Areas of the topology which have not been validated to evaluate the topology rules are referred to as "dirty areas." Once a dirty area is validated, any errors discovered are stored in the database. These errors can be fixed using editing tools in the GIS, they can be left in the database as errors, or they can be marked as exceptions to the rule. In today's very large databases, this gives considerable flexibility to workflow rather than having to validate an entire coverage before usage.

The editing tools for fixing errors have broad applications. Editing can be done on two feature classes at once. Features can easily be merged and split. Features in one class can be constructed based on the geometry of selected features in another class.

In summary, topological editing tools enforce spatial integrity constraints, across multiple feature classes. The arrangement whereby the rules are established in a separate data set rather than being embedded in the data gives a workflow flexibility within an organization.

The applications of these capabilities are substantial. Imagine, for example, incorporating new construction that changes the location of an associated road. The topological differences between the road centerline and the school district can then be identified and resolved. Now imagine this on a much larger scale. In one Canadian city they desired to integrate all the features mapped on a citywide CAD system into the GIS system for the same city, a potentially expensive process. The ability to use multilayered topology greatly facilitated this process and reduced the expense considerably. Topology in the geodatabase is a desirable tool to ensure accurate data in all layers. It is continually available as features and layers and relationships are added or amended in the GIS.

Temporal data

GIS have typically been poor at handling temporal data in the past, at best providing multiple static overlays. This is changing with the development of software that can keep track of events that take place at different times in either the same place or at different places.

A simple event includes the ID of the object, the time, the place and, if necessary, the status of the object. This places a dot in space at a time in a certain condition. Several simple events can be linked to form a track.

Complex events can be handled. These have additional information about the nature of the object being tracked. A dynamic complex event could track a plane of a certain model, flight number, number of passengers, fuel load, age of aircraft and name of pilot. Complex stationary events can be envisaged resulting from the use of traffic sensors.

Currently the software handles moving points or points over time. In the immediate future, points, lines, and polygons will be able to be handled in a similar manner. Examples of lines are military fronts or weather fronts. Polygons might be satellite footprints or oil slicks or temperature maps or precipitation maps for example.

The software allows for the mapping of these temporal events in geographic information systems. The data can be dynamically presented and be played forwards or backwards. Real-time or near real-time tracking of data is possible. Data rates will depend on the communication links available, server speeds, and network speeds. Real-time tracking may include emergency response systems, threat detection, fleet tracking, or satellite tracking systems. A wide range of symbology can be employed which can change as the status of the event changes (as a hurricane intensifies, for example). With each playback a histogram is available showing the number of events occurring over time and this can be used for analysis of the events themselves (e.g., frequency of enemy shelling) or can be used to determine the playback time or repeated playback times for further analysis. All of the displays can be animated for use on a media player, and have plug-and-play capability with your own animation engine.

It is possible to show single or multiple temporal events in the form of a clock. The clock wizard creates a circular chart, or clock, of temporal data that can be used to analyze patterns of the data that may be missed when viewing the data on a table or map. The clock bands can be modified, colors can be changed, classes and legend settings can be used to produce temporal data that illustrates specific aspects of the analysis, such as the nocturnal patterns of animal movement.

Users can also create and apply preset or custom actions and query temporal data based on location information, feature attributes, or a combination of the two.

Cartography

While GIS can produce elegant and beautiful maps, the process has often been cumbersome, particularly when multiple maps had to be created. This state of affairs is rapidly changing.

It is now possible for multiple maps to be made from a single geodatabase with a consistency of content and feature selection. This allows frequent updates to be cost effective, as they are done once rather than on numerous files. Automation can be achieved in the higher qualities of map sophistication rather than only in the lower mass-produced cartography. This use of intelligent cartography rather than "brute-force" cartography will be applied to large-scale mapping operations. Sophisticated allocations of symbology can be anticipated. Take for example the conceptually straightforward task of legend placement. The rule could be "place a legend somewhere over water or white space on the page such that it is no closer than 1.2 cm from either a coastline, the edge of the page, or another element." Such a task is easy to define even now. However, mapping applications do not yet work with sufficiently refined notions of the shape of an object, or have the ability intelligently to move objects like legends relative to coastlines. The main issue is the quality of the result; as computing power and as the ability of displacement algorithms improve with regard to displacing complex geometries against other complex geometries, the quality is improving.

Process models for intelligent map creation have not been especially well defined as a process or as a structure for handling the process.

For example, to "construct a base map from a variety of GIS data layers" a mapmaker must know what procedures are needed for each layer in the map, which features are to appear, how to symbolize them, how text is placed for them, and what the priorities are for displaying the features (their relative importance). As problems go, this is complex and has several dimensions that must simultaneously be accounted for. Assigning a sequence to a set of operations, based on the interdependent relationships that exist between the data both logically and spatially, that would be used to construct the map, could create a process model for a map. This model, in an interactive software modeling environment, would allow a mapmaker effectively to match data sources to layers, which are then used as inputs to the model, and then "run" the model. Such a model is modestly difficult to construct to work for one edition of a given map, but in a subsequent edition things may have changed and therefore the model must be flexible enough to evaluate the data and then make the appropriate choices in processing the data or constructing a map.

Automated page layout, text placement, generalization, and publishing continue to grow and to evolve.

Automation of generalization will be possible if models can be created to move from one known map design to a second known map design at a different scale. Given the "bookends," the "row of books" can be designed and modified to suit varying cartographic designs in terms of sophistication. Such a solution depends heavily on a database design that is intended to support the processing needed to produce each "book." Meanwhile, the intelligent cartographic production of high-quality maps from a single geodatabase is becoming available for a growing array of map types in many industries.

Network analysis

Taking advantage of the object-relational data model (see chapter 9), a new data structure for networks in geodatabases is being developed which will allow for much more realistic modeling of network connectivity for the purposes of routing and tracing. This data structure fully supports multimodal networks, such as found in transportation (car, bus, bike, rail, shipping, air routes) and hydrology (stream channels, roadways, wastewater, basins). Network connectivity can be based on geometry using a very rich connecting model, as well as database relationships, such as air-flight relationships between airports, or bus-route relationships to bus stops. These database relationships can be thought of as "virtual pathways," and allow for multimodal routing, where people, commodities, or ideas are transported by road, rail, ship, or airplane, transferring between each mode at known transfer points.

Turns are modeled as features, allowing them to be conflated between different feature classes. Turns can also be modeled as "implicit" turning movements, taking into account turning angles.

Networks are made of junction, edge, and turn elements, derived from the geodatabase feature classes and object relationships. Elements have any number of attributes, such as travel time for cars, buses, emergency vehicles, heavy trucks, costs, restrictions, slope, number of lanes, pipe diameter, and so on. These attributes can be calculated from field values found in geodatabase tables or from a script. Attributes can be "dynamic," meaning they are calculated on demand. A dynamic attribute might query a recycling data structure containing current traffic conditions downloaded from the Web.

Network data sets are versioned, to support planning requirements and multi-user editing. Networks are incrementally built, meaning that only edited portions of the network are rebuilt, rather than requiring a full rebuild on each edit.

The true value of the network structure is that it allows a flexible and extended query capability for the production of information products based on the network. These capabilities include:

- Shortest path. Find the least-cost path (time, distance, etc.) through a series of stops. (Stops are network point locations)
- Closest facility. Given one location, find the closest (time, distance, etc.) facility (ATM, hospital, Starbucks, etc.)
- Traveling salesman. First find the optimal sequence to visit set of stops, then run a shortest path.
- Allocate. Build a shortest-path tree to define a "service area" or a "space-time defined space."
- Origin/destination matrix (OD Matrix). Build a matrix of cost between a set of origins (O) and destinations (D). OD Matrixes are used extensively in many network analyses (such as Tour, Location/ allocation, etc.). Matrixes are represented as a network.
- Vehicle routing. The core route optimizer. Given a fleet of heterogeneous vehicles, in terms of capacity, available hours of service, cost to deploy, overtime costs, etc., and a set of heterogeneous customers, both vehicles and customers with time window constraints, find the optimal vehicle/customer assignment and route for the vehicle.
- Location/allocation. Simultaneously locate facilities and assign (allocate) demand to the facilities.
- "Chinese postman." Find the optimal path through a set of connected edges. Garbage trucks, newspaper delivery.
- Tracing. Works on a directed network, where "flows" in one direction, such as water, electrons, wastewater, etc., can be handled. Tracing tasks can be thought of as a general query: select what's upstream,

downstream, find cycles, trace both up and downstream, or find dangles.

Data conversion requirements

At this point, it is necessary to examine the conversion processes that will be used to get information from different sources into your GIS. There are a variety of methods available. The one you choose will depend on the format and quality of existing data, the format of data from outside sources, and the standards you have set. Your basic options are to develop the data in-house, have the data prepared by an outside contractor, or reformat some existing digital data.

In-house database development is typically accomplished using one or more of the following methods:

- Digitizing
- Scanning
- Keyboard input
- File input
- File transfer

Using data prepared by an outsource contractor is common in situations where the database must be developed rapidly, there is no or little in-house capability, or ongoing maintenance and further database development are not anticipated. When using an outside contractor for this crucial step (remember the "garbage in, garbage out" rule) take a close look at past experience with the vendor, his familiarity with the software and hardware you are using, and the availability of updates. Other GIS users with similar types of applications and database requirements are good sources of information on commercial vendors.

As the sources of digital spatial data proliferate, the practice of reformatting existing digital data into a format usable by your GIS is becoming common. There are about 30 digital data exchange formats currently in use, including digital line graph (.DLG), the U. S. Census Bureau's TIGER® (which stands for Topologically Integrated Geographic Encoding and Reference system), and the shapefile, ESRI's published exchange format. An interchange format to get data from computer-aided design (CAD) systems is called .DXF. When reformatting data, you should consider the system you're exchanging data with and the digital interchange format that's necessary.

Each data conversion method takes time and has an associated cost. The development of digital data exchange formats and the proliferation of conversion and translation software have made sharing data easier, but you may still encounter problems reformatting data from other systems. Don't assume that the data is easy to use because it is in a digital form. There are times when you find that the cost of reformatting data into something usable is not cost-effective. For example, polygon data from one type of software may only be graphic with no associated attributes or topology. The cost to make this data suitable for your work may be greater than the cost of in-house digitizing from hard copy. Another potential difficulty is that you may not get documentation that is adequate for determining the accuracy, currency, source, or anything else about the data, which means you'll have no way to verify that the digitized data will meet your needs. Such data can only be used with extreme caution and depends solely on the credibility of the source (if known) and how desperate the need for the data. A useful way to think of the integrity of data is: 80 percent of data = 20 percent of problems, 20 percent of data = 80 percent of problems.

Chapter nine

Choose a logical data model

"The new generation of object-oriented data models are ushering in a host of new GIS capabilities, and should be considered for all new implementations. Yet the relational model is still prevalent, and the savvy GIS manager will be conversant in both."

Choose a logical data model

At this point in the GIS planning process you should have a clear understanding of the data elements that are needed to produce your information products and you should also have identified by name any logical links that will be required between the various data elements. You also understand the limitations and dimensions of your data, and where your data is going to come from.

The next step is to describe the actual structure of the data using one of several logical data models: relational, object-oriented, or object-relational. The system's end users aren't directly concerned with this issue. But for you, the GIS leader, it becomes a significant concern at this juncture because the organization of the database will factor into the decision of which software system ultimately gets deployed. The crux of the issue is that a logical data model must describe a complex version of the real world in a database. Determining the best logical model and then building it requires a decision and now is the time to start thinking about it. How much it will cost to build and use such a database will depend on how closely you want to model reality.

Let's examine the three types of logical data models.

The relational data model

The vast majority of the geospatial digital data in the world currently is stored using relational data models. In a database set up as a relational data model, the data are stored as collections of tables (called relations) that are logically associated to each other by shared attributes. The individual records are stored as rows in the tables while the attributes are stored as columns. Each column can contain attribute data of only one kind, date, character string, numeric, and so forth. Tables are typically normalized (calibrated to one another) to minimize redundancy. Storing spatial data in a relational database is greatly facilitated by using specialized software such as Spatial Database Engine™ (SDE®) by ESRI that allows the GIS to read spatial data stored in common RDBMS such as Oracle®.

Your GIS links spatial data to tabular data. In the example below, you can see the links between the spatial data displayed on the map and the table containing attributes about features on the map (e.g., parcel number). A separate but related table uses the common Parcel-No attribute column to link to another table showing owner information. There could be many additional tables available for use in GIS analysis and thematic display all linked back to one another by a thread of common fields.

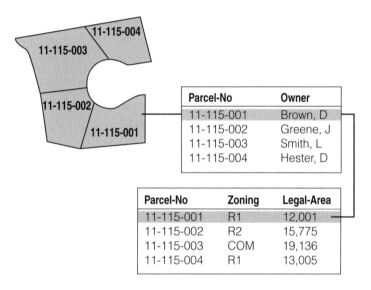

Relational data models use a fixed set of built-in data types such as numbers, dates, and character strings to segregate different types of attribute data. These offer a tightly constrained but highly efficient method of describing the real world. They execute and deliver results quickly, but require sophisticated application programming to model complex real-world situations in a meaningful way.

Consider the task of dispatching an emergency vehicle. Finding the quickest route requires detailed information about the direction and capacity of each street at different times of the day and the types of signals and change controls at each intersection. These complex variables would require a large number of interacting tables. The strength of relational tables is that they simplify the real world and give swift and reliable answers to the queries that they can handle.

Components of the relational data model

When you develop your logical data model using the relational approach, you may need to consider the layering and tiling structure of your data, also known as the map library. (There is a newer approach that uses seamless databases, but it doesn't work under the relational model. For that you need an object-oriented approach.) A relational library of map partitions organizes geographic data into data sets of manageable size in an intuitively navigable tile system.

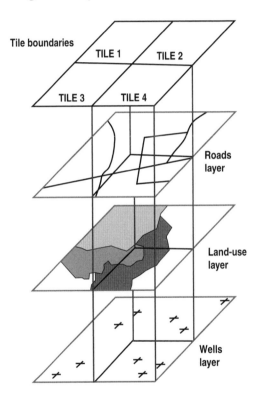

The library structure you choose is important; it will affect database maintenance, data query, and overall performance. At this point, you should finalize the layer-naming convention and the tile structure.

Review again the master input data list to compile your final list of all the necessary layers. Settle on a unique (hopefully descriptive) name for each and every layer. Once agreed upon by the GIS team and published to the entire user group, these layer names are to be adhered to rigorously throughout the rest of the conceptual database design. It's good practice to name layers descriptively so that the contents can be easily recognized.

The tile structure is the spatial index to the data in your GIS. Once established, it is labor-intensive to change, so think carefully about your tile selection. For example, it is better to frame your tiles using physical objects such as roads than on political boundaries that may change over time. Also, an abstract grid such as USGS quadrangle boundaries offers a stable, somewhat standard, tile structure. Tiling speeds up access because the system indexes the data geographically allowing you to search just the area of immediate interest and not the whole thing. Remember, GIS data sets tend to be large, so anything we can do to cut to the chase pays dividends. If your information product descriptions show that your most important applications will need access by quadrangle, then making the same quadrangles your tiling unit could be beneficial.

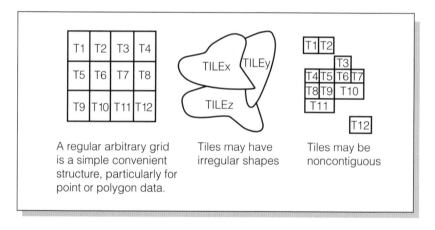

A regular arbitrary grid is a simple convenient structure, particularly for point or polygon data.

Tiles may have irregular shapes

Tiles may be noncontiguous

Map projection

Use a single coordinate system for all the data layers in a map library. Data can originate from different projections, but should be projected into a common coordinate system before being added to the library. This is imperative, since once you're in a mapping display mode, data with different coordinate systems or projections will not even appear in the same space.

Units (imperial or metric)

Use a single set of map units for your library. Sometimes this is determined by the map projection and coordinate system you choose. For example, the State Plane coordinate system typically stores data in feet, while UTM stores data in meters.

Precision (single or double)

The storage precision of x, y coordinate data is important. Review your information product descriptions to determine what precision is necessary for your organization. Coordinates are either single-precision real numbers (six to seven significant digits) or double-precision (13 to 14 significant digits). The precision you choose also impacts your data storage requirements, because, as you might expect, double-precision requires more data storage capacity.

Layer

A layer is a logical grouping of geographic features that can also be referred to as a coverage or theme (coverage and theme are the terms native to ESRI ArcInfo and ArcView programs). You should determine the content of each data layer as part of the conceptual database design. How you organize your data layers will depend on how you will be using the data. Much of this thinking was done when you prepared the master input data list. At this point, review the list and make any necessary modifications.

Features (entities)

Normally, layers are organized so that points, lines, and polygons are stored in separate layers. For example, parcels represented by polygons might be stored in one layer, while roads represented by lines might be stored in another layer, while hydrants stored as points might be stored in yet another. (And just to spice things up, these days we also have cell-based grids representing images.)

Additionally, features should be organized thematically. For example, roads and streams could both be represented as lines, but it wouldn't make sense to represent them on the same line layer because they are different things and therefore need to be considered independently.

Attributes

Typically, you will identify a set of attributes needed for the features in each layer. For example, for each well site (a point) you may want to know the well-identification number, the depth, the pipe diameter, type of pump, and gallons-per-minute. These are the data elements required for your information product, so obviously these attribute columns will need to be present in the layers.

Intended uses

You also need to know the intended use of the data. For example, if your city's water source comes from a series of public water wells, you should determine if you want only public wells in this layer or if you also want to include private wells. It is important to review your information product descriptions to determine the data requirements for each layer.

Logical linkages

When designing your layers, make sure that logical linkages exist between the data layers and any attribute files necessary to make your information products. It is important to look for logical linkages that might not be there. If you identify logical linkages that are needed to make an information product but are not currently in the data, you have a task that needs to be completed. A common field, say Parcel-No, is referred to as a key if it is contained in both tables and can be used to link them together.

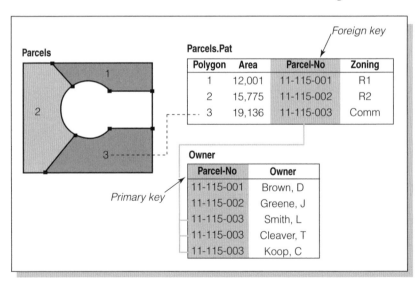

Logical linkages between parcels and their owners

Source

It is important to know and document in the metadata the source of each data layer. The source information will affect the data standards set for each map library.

Data accuracy and standards

Data standards ensure that the data layers in your database will serve as source data for meaningful analysis. For example, earlier we mentioned that all layers in a map library should have a similar resolution. Think of what would happen if you forgot that the resolution of a certain input data set was only 1:200,000 while all the other data sets in that map library were digitized at a much higher resolution of say 1:6,000. Using the lower resolution data set in an overlay analysis with the rest of the data could have a profound impact on the validity of the output results. Worse yet, the users might not even realize there was a data accuracy issue and might base a crucial business decision on erroneous analysis. Making sure this can't happen is incumbent on the GIS leader, and the way to make sure is through rigorous documentation in the metadata.

The following matrix shows a list of some data layers from a typical municipal government GIS. It shows the layer name, the real-world objects represented by the name, the type of feature, and the required attributes for each layer. You need to develop a matrix, probably much longer than this one that describes all of the layers in your design. Keep in mind that it's not unusual for layer counts in an established GIS in a large municipal setting to number in the hundreds.

Layer name	Real-world object	Feature type	User attributes
Street	City streets	Lines	Name; class of street
Blockgroup	Census blockgroups	Polygons	Age group population counts; median household income
Zone	Land-use zoning	Polygons	Zoning type
Land use	Actual activity of land use	Polygons	Activity type
Railroads	Main train lines	Lines	Railroad name
Sewer complaints	Incident locations	Points	Address, date, and description of incident

Advantages of the relational data model

There are a number of general advantages to the relational model including:

1. simple table structures which are easy to read
2. intuitive, simple user interface
3. many end-user tools (i.e., macros and scripts) are available
4. easy modification and addition of new relationships, data, and records
5. tables describing geographic features with common attributes are easily used
6. attribute tables can be linked to tables describing the topology necessary for a GIS
7. direct access to data provides fast and efficient performance.
8. independence of data from the application
9. optimized for GIS query and analysis
10. large amounts of GIS data are available in this format
11. large pool of experienced developers, developer tools, textbooks, and consultants

Disadvantages of the relational data model

As you might expect, there are also a number of disadvantages to the relational model including:

1. limited representation of the real world
2. limited flexibility of queries and data management.
3. slow sequential access
4. complex data relationships are difficult to model often requiring specialized database application programmers
5. complex relationships must be expressed as procedures in every program that accesses the database
6. performance penalty due to the need to reassemble data structures every time the data is accessed

The object-oriented data model

Newer than their relational counterparts, object-oriented data models allow for rich and complex descriptions of the real world and the ability to set up a data structure that users will find easy to understand. Objects can be modeled after real-world entities (like sewers, fires, forests, building owners) and can be given behavior that mimics or models some relevant aspect of their behavior in the real world. One simplistic way of thinking about the difference is that the object stores the information about itself (all its attributes) within itself instead of in a bunch of related tables.

An object model of a street network, for instance, would use lines depicting streets and each street segment shows behavior that models its real-world actions, such as the directions of traffic or the amount of cars per hour that can traverse it at any given hour of the day.

Objects
Objects represent real-world entities such as buildings, streams, or bank accounts. Objects have properties that define their state and methods that define their behavior. Objects interact with one another by passing messages that invoke their behaviors.

Attributes
Attributes are the properties that define the state of an object, like the speed limit of a street, the name of a building's owner, or the peak capacity of a storm drain.

Behaviors
Behaviors are the methods, or the operations, that an object can perform. For example, a street may know how to calculate the increase in time needed to travel its length as rush hour approaches, or an account may know how to subtract money from its balance when a withdrawal is made. These behaviors can also be used to send messages to other objects, to communicate the state of an object by reporting current values, to store new values, or to perform calculations.

Messages
Objects communicate with one another through messages. Messages are the act of one object invoking another object's behavior. A message is the

name of an object followed by the name of a behavior the object knows how to carry out. The object that initiates a message is called the sender while the object that receives the message is called the receiver.

Classes

A class is a way to group objects that share the same set of attributes and behaviors into a template. Objects of a particular class are referred to as instances of that class. For example, the land parcel where you live is just one of many parcels that exist in your city, and each is unique. But these parcels also share certain useful characteristics, like building type or zoning code, and these shared characteristics are expressed as classes.

Classes can be nested to any degree, and inheritance (see the following) will automatically accumulate down though all the levels. The resulting tree-like structure is known as a class hierarchy. Determining the classes that you will need is an important step in your database design.

Relationships

Relationships describe how objects are associated with each other. They define rules for creating, modifying, and removing objects. There are several kinds of relationships that can be used in the object-oriented data model. These are:

- Inheritance allows one class to inherit the attributes and behaviors of one or more other classes. The class that inherits attributes and behaviors is known as the subclass. The parent class is referred to as the superclass. In addition to the behaviors they inherit, subclasses may add or override inherited attributes and behaviors. A superclass is referred to as a generalization of its subclasses and a subclass is a specialization of its superclass. For example, a house is a specialization of a building and a building is a generalization of a house. A house class can inherit attributes and behaviors of the building class such as number of floors, rooms, and construction type.

- An association is a general relationship between objects. Each association can also have a multiplicity associated with it that defines the number of objects that are associated with another object. For example, an association might tell you that the object "owner" can own one or many houses. Aggregation and composition are specialized types of associations.

- Aggregation is a particular type of association. Objects can contain other objects, so aggregation is simply a collection of different object classes assembled into an aggregate class, which becomes a new object. These new composite objects are important because they can represent more complex structures than can simple objects. For example, a building object can be aggregated with a sign object. But if you delete the sign, the building still remains.
- Composition is another specialized form of association. This is a stronger association relationship, in which the life of the contained object classes controls the life of the container object class. For example, a building is composed of a foundation, walls, and a roof. If you delete the building, you automatically delete its foundation, walls, and roof, but not its sign.

Encapsulation of behavior

Encapsulation is the essence of the object-oriented model. In an object-oriented model objects encapsulate attributes and behaviors. The data within an object can be accessed only in accordance with the object's behaviors. In this way encapsulation protects data from corruption by other objects and also masks the internal details of objects from the rest of the system. Encapsulation also provides a degree of data independence, so that objects that send or receive messages do not need to be modified when interacting with an object whose behavior has changed. This allows changes in a program without the cost of a major restructuring, as would be the case with a relational structure.

Class diagrams

Class diagrams are used to diagram the conceptual database design. They help you map the relationships that you need in a data model. They are used to illustrate the classes and relationships in your database. The following class diagrams illustrate how to diagram classes, attributes, methods, and their relationships. The diagrams are a standard notation for expressing object models and are based on the Unified Modeling Language™ (UML™), the emerging standard.

139

Recall that a class is a set of similar objects. Each object in a class has the same set of attributes and methods. In the example below, the class named Parcel has attributes of feature-polygon, size, cost, and zoning, and can perform these methods: calculate tax value, split parcels, and merge parcels.

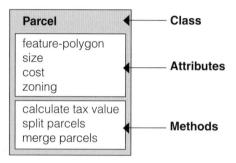

Inheritance or generalization is the ability to share object properties and methods with a class or superclass. Inheritance creates a new object class by modifying an existing class.

Associations represent relationships between classes. Aggregation and composition (discussed in the following) are specialized types of associations.

Multiplicity of associations

Multiplicity defines the number of objects that can be associated with another object. In the following diagram we see that a water valve can have exactly one documentation reference and vice versa. In contrast, a pump station can have many pumps associated with it, but each pump can have only one pump station. While this might seem obvious, the rule system that this suggests is what gives object-oriented modeling its strength. Consider the next example: a pole can have zero or one transformers, but a transformer can have only one pole. In the final example, we see that an owner can own one or many land parcels, and a land parcel can be owned by one or many owners.

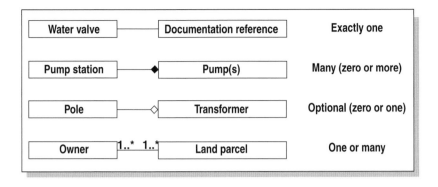

Water valve — Documentation reference		Exactly one
Pump station —◆ Pump(s)		Many (zero or more)
Pole —◇ Transformer		Optional (zero or one)
Owner 1..* 1..* Land parcel		One or many

Aggregation

Aggregation is an asymmetric association in which an object from one class is considered to be a "whole" and objects from another class are considered to be "parts." Object classes can be assembled to create an aggregate class. For example, the aggregate class Property can be created by aggregating the Land parcel class and the Dwellings class.

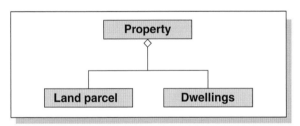

Composition is a stronger form of aggregation in which objects from the "whole" class control the lifetime of objects from the "subordinate" class. If the object that makes up the "whole" is deleted, the "subordinate" objects that compose the whole are also deleted.

In the following examples, you have a Network class of water features, with subordinate classes of Stream and Canal. When a Network is deleted, the Stream and Canal objects that compose it are also deleted.

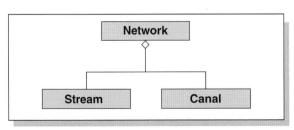

Chapter nine

There are a number of advantages to the object-oriented data model:

1. Encapsulation combines object attributes and behaviors making an object accessible through a well-defined set of methods and attributes so that you don't need to know the inner workings of an object.
2. Allows complex representations of the real world.
3. Supports multiple levels of generalization, aggregation, and association.
4. Maintains history in the database.
5. Integrates well with simulation modeling techniques.
6. Multiple simultaneous updating (versioning).
7. Intuitive feel because it uses objects that occur naturally.
8. Well-suited for modeling complex data relationships.
9. Requires less code in GIS programs, meaning fewer bugs and lower maintenance costs.
10. Ensures a high level of data integrity (new data must follow the behavior rules).

There are also some disadvantages to the object-oriented data model. These include:

1. Although object-oriented data models allow complex representation of the real world, complex models are more difficult to design and build. The choice of objects is crucial.
2. Import and exchange with other types of databases is difficult.
3. Some business applications may not be able to access or contribute to an object-oriented database.
4. Large and complex models are slow to execute.
5. Dependent on a thorough description of real-world phenomena (particularly difficult in the natural world).
6. Object-oriented databases require the use of object-oriented computer languages for their analysis. Fewer people are trained in object-oriented programming.

The object-relational data model

The most recent development in the domain of logical data models is the object-relational model. These systems build object-oriented capabilities on top of a robust relational database. The relational database is extended by software that incorporates object-oriented behaviors, but data are not encapsulated. Database information is still in tables, but some of the attribute columns can include a richer data structure, called an abstract data type.

The object-relational model has advantages of speed (important in large databases) and the ability to handle complexity as well as the database-building integrity of object-oriented designs. It has the additional advantage of supporting an extended form of Structured Query Language (SQL) and can access typical RDBMS (relational databases). This alone may be an important consideration in enterprise-wide systems where other business applications need to access or contribute data to the GIS database.

Object-relational data models incorporate characteristics of both the relational and object-oriented databases. Recall that the relational model uses tables with a fixed set of built-in data types (e.g., numbers, date), while in the object-oriented data model, the objects have unique attributes and behaviors are encapsulated within the object.

The object-relational model extends the relational model by adding a new type of data structure called the abstract data type. Abstract data types are created by combining the basic alphanumeric data types used in the relational model. The object-relational model allows you to add specialized behavior to the relational model. This flexibility allows the object-relational model to more closely model the real world than can a pure relational database model.

The extent to which the object-oriented data model component can be implemented in the object-relational model is rapidly changing. Object-relational software is continuing to improve and add more object functionality.

There are a number of advantages to the object-relational data model:

1. Fast execution.
2. Uniform repository of geographic data; allows use of legacy and non-GIS databases.
3. Data entry and editing are more accurate.
4. Data integrity is high (new data must follow the behavior rules).
5. Users can work with more intuitive data objects.
6. Simultaneous data editing (versioning).
7. Less need for programming applications to model complex relationships.

Of course there are a number of disadvantages to the object-relational data model:

1. Compromise between object-oriented and relational data models.
2. No data encapsulation.
3. Limited support for object relationships.
4. Complex relationships are more difficult to model than when using a pure object-oriented data model.

ESRI ArcInfo 8.x technology is a well-known example of an relational model extended to optionally include object-oriented behavior in a data model called the geodatabase data model. The geodatabase has several features that take advantage of the object-relational data model concept including attribute domains, network connectivity rules, and relationship rules.

The types of functions required to create your information products are invaluable in helping you determine what data model is appropriate for your GIS. For example, if you needed to perform a lot of network analysis with complex intersections or junctions, you may choose an object-relational model. If your information products do not require the type of functionality provided by object-oriented capabilities, you might want to choose the more traditional relational data model. Choosing which logical data model is appropriate for your organization is not always a straightforward task. Much is being said about the object-oriented and object-relational data models and how they will impact the traditional relational implementation of GIS. However, this decision should be based on more than the conceptual differences.

Since the 1980s, most GIS databases have been based on the relational data model. But given the difficulty faced by this model in supporting the complex behaviors of real-world objects, some GIS software vendors have utilized the object-oriented and object-relational models to support these more complex data models. Choosing which model is most appropriate for your organization will in part determine which software you purchase. If a primary need is to model complex data relationships, then you better well have a system that supports it. It boils down to the question of which data model is most suitable for the work you want to do.

The following table describes some typical data modeling situations along with the logical model characteristics that are needed, and suggests a logical data model that would be appropriate.

Data modeling situation	Logical model characteristics	Suggested logical data model
Forest inventory of tree stands, rivers, and roads for forest-harvesting analysis.	Simple relationships between features.	Relational
Addition of new forest stand attributes as forest matures. Addition of features such as property boundaries.	Easy modification and addition of new features and attributes as time passes.	Relational
Small staff. Need to minimize training and implementation time.	Simple, easy-to-use interface. Database that is simple and easy to design and build.	Relational
Enterprise-wide system must connect to existing sales and business partner databases.	Connects well to existing databases.	Relational
Large business needs to perform suitable site analysis for locating new stores.	Need to use existing demographic data.	Relational
Need to do real-time flood forecasting along a river system.	Complex representation of real world.	Object-oriented
Need to simulate traffic flow through a street network during an emergency situation.	Integration with sophisticated simulation models.	Object-oriented
Major utility company needs to simultaneously update many parts of its large database as daily additions and repairs are made.	Multiple simultaneous updating (versioning).	Object-oriented or object-relational
Database is in constant use in mission-critical life-dependent situations.	High level of data integrity.	Object-oriented or object-relational
Numerous new applications will be developed over time.	Low application development costs once initial model is developed.	Object-oriented
Complex analysis of natural resource features in large watershed (e.g., Columbia River Basin).	Fast to execute, particularly for large, complex analysis.	Object-relational
Many legacy relational databases and non-GIS databases to be linked to new GIS.	Links well with all types of databases.	Object-relational
Water utility needs to model water network including water mains, laterals, valves, pump stations, and drains.	Complex relationships. Inheritance of attributes and behaviors.	Object-oriented or object-relational
Large amount of data maintenance and update.	High level of data integrity.	Object-oriented or object-relational

In addition, to determine which logical data model is most appropriate for your needs you can compare the assumed cost for developing your priority information products under each data model. A general approach to use to select a logical data model is to compare costs for your high-priority information products. The costs can be assigned to database creation and application programming categories. Database costs include the cost for developing and maintaining the database(s) needed to create the information products (e.g., database design, data conversion, cost of ongoing data exchange with legacy systems, staff training, and so on). Application costs include the cost of programming the application and any database manipulation required specifically for the information products. This approach can be used with actual figures if that data is available, or using relative rankings if the data is not available.

If you look at the real costs of preparing 20 or 30 information products over the life of your system using the different logical data models, you can conclude which data model is most appropriate for your organization.

Determine system requirements

"Getting the system requirements right at the onset is what separates the men from the boys in successful GIS planning."

Determine system requirements

The next stage in the GIS planning process is to design a technology system that can handle the data and functionality you've specified. The equivalent conceptual system design for your data (which you learned about in the chapter 9) and the conceptual system design for technology featured in this chapter, are presented as separate steps, but they are not necessarily sequential tasks. They are typically carried out at roughly the same time. The conceptual system design for technology is focused on defining a set of technologies (hardware, software, networking) that will adequately support the demand for system functions as the needed information products are created. This is also the place to consider the implications of distributed GIS and Web services.

This chapter will guide you through the process of planning the system technology and the important step of documenting your design concept. At the conclusion of the chapter you'll see how all the components discussed up to this point in the book (IPD, MIDL, data design, technology) should be arranged into a cohesive and understandable preliminary design document.

Function requirements: summary and classification

In chapter 6 you determined the functions required to get data into the system and to generate each information product identified. Now it's time to summarize and classify these functions. Doing this will allow you to understand the total function utilization of your system. You can buy the latest hardware and software available, but money and resources are wasted if they are missing a key bit of required functionality (i.e., you choose a software program that does not easily import the data from the public works department where you are planning to get some crucial data). Equally critical to your ultimate success are the speed and efficiency that you'll be able to execute these functions.

Chapter ten

Summarizing the function requirements

The functions required for your system were already identified and are listed in the individual information product descriptions and also on the master input data list. Your job at this point is simple: review the IPDs and summarize the number of times that each specific function will be used to produce the full set of information products. Add to this summary the additional data input functions identified in the MIDL (things like data import and conversion functions). The result is called the "total function utilization."

Be sure to forecast ahead sufficiently and include every function that you can see invoked at any time over the next set number of years. For the purposes of this description we will use five years as the planning period. There are two sources of information already established for you to make this calculation:

In chapter 6 we showed how to calculate the number of times each function was used to make each individual information product over a five-year span. This took into account the frequency with which each information product was to be made each year. Adding the totals for each year gave us the total function requirements by function.

We then summarized all the functions we'd need to get data processed plus any other basic system capabilities. For all of the data sets that will be input in the first five years (typically all of them), total the number of times each function will be used.

Now we combine the estimates of function utilization in a simple table, function name on the left side, total number of times function is invoked during planning on the right hand side. Rank this table by frequency of use, starting with the most frequently used functions. This is your total function utilization table. It is sometimes helpful to produce a simple graph from this table with the ranked functions in order along the bottom and the total number of times the function is used on the side. This is the total function utilization graph.

As an optional step, you can calculate the proposed use of each function on a year-by-year basis. This could be useful if you require limited functionality in the early years of a system and a higher level of functionality as you went along in years. You have already calculated the yearly function requirements for each information product and for all data sets.

150

With this data, it is relatively straightforward to take it one step further and determine the functions required on an annual basis.

Classifying system functions

It is useful to classify system functions in other ways besides sheer volume to provide a better understanding of the importance of factoring function requirements into the technology equation. Looking at the graph of total function utilization, the first thing that pops out is that the functions that carry out basic system chores (like data input) are used frequently and thus are essential to your system's operation.

Typically there are one or a small group of data manipulation and analysis functions that have an extremely high frequency of use. Clearly, these functions are important: your system relies heavily on them. These are classified as class 1 functions. In the software you ultimately select, these are the functions that must be proven not just to be present, but to be optimally suited and operationally efficient. Any system that does not offer suitable class 1 functions should be automatically disqualified from consideration.

It is easy to recognize the least-used functions. There usually are quite a number of them along the bottom of the graph on the right-hand side. These are the functions that need to be in place in any system acquired. They are not frequently required, but may be critical to individual operations. They need to be present, but not necessarily efficient. These are called the class 3 functions.

The remaining functions in the middle of your graph provide essential functionality and are heavily used. These are class 2 functions. When benchmark testing systems prior to a major procurement, class 2 functions are always thoroughly tested. They must be in place and they must be efficient. The following example graph identifies the classifications of functions.

Document the classifications of your required system functions. The document should include the functions required by the GIS, the frequency of their use, and the functional classification. This will be included in the documentation provided to vendors when you procure your system. Note that no other functions need enter into your procurement. This will help to keep you focused.

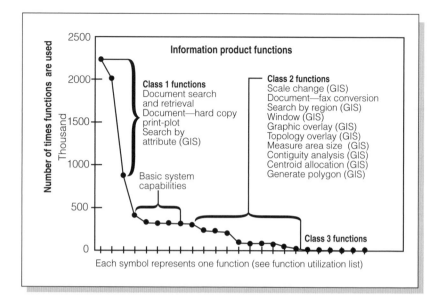

During the classification process, keep in mind that the numbers you are working with are estimates and that you should expect variations in the methodology and information product timing over the five-year plan. At this point in the planning process, you can figure your estimates to be off by 20 or 30 percent. The objective is to ensure that your technology design can accommodate a plus or minus 30 percent change in functional requirements, and not be 200 or 300 percent wrong.

Interface and communication technologies

Determining an adequate system interface and network communication configuration requires an understanding of where the major data sets in yours and other organizations live, where they're stored, and how you're connected to it. These data sets were identified in the master input data list (MIDL).

It is likely that you'll need to link many of your organization's existing databases to your GIS. If your GIS requires frequent, repeated, high-speed access to major databases, this needs to be specified. Ideally these are two-way links: you're getting data and sending different data. Consider the city information systems folks who want to link their utility billing records to the GIS parcel database. If the two databases are stored in different formats, then special interface software (i.e., a system interface) might be required to allow for this linkage. This requires programming, which costs time and money.

The data records involved, and the frequency and speed of access are the determining factors for choosing a system interface. Sometimes, all this discussion of database technology makes organizations realize they are using dated technology and triggers an organization-wide migration to a new standard database platform on which all applications, including the GIS, will successfully operate. If special interface software is needed, estimate the programming cost involved.

Network connections provide the communications link that allows you to access distributed data throughout your organization. A typical municipality, for example, has a centralized server that stores data to which the various departments have access. Most city employees have access to the data on the server through their desktop computers. It is the network and its communication protocols that allow this communication and data sharing to occur. The GIS may use the existing network to provide data access to GIS users, or it could require updating or the adoption of an entirely new networking system. The sheer bulk of GIS data is enough to choke undersized network connections. Couple high data loads with frequently used applications, and you've got a recipe for bottlenecks right out of the gate. Don't be fooled: your GIS will have an impact on an existing network, especially once you're successful and growing, so always anticipate the highest possible numbers and then add 20 percent to be safe. The existing network may require significant changes to accommodate the demands produced from GIS implementation.

There are two basic network types: local area networks (LANs) and wide area networks (WANs). LANs support high-bandwidth communications over short distances. They provide high-speed access to data, typically within a building or other localized environment.

WANs support communications between remote locations. A city office building, for example, may have a file server that shares data through a WAN to its field offices. WAN technology typically supports lower bandwidth than LAN environments, but data transmission is possible over long distances. The Internet, in essence, is a global WAN. Due to the cost of constructing the communications infrastructure for a WAN, its costs are greater than those of a LAN. (The implications of Internet-distributed GIS is addressed later in this chapter.)

153

A server is capable of storing large amounts of data, but like your own desktop computer, the server can hold only as much data as its disk space permits. A small municipality may require only 20 to 40 gigabytes of disk space to meet its data storage requirements. Large data repositories (like Alexandria Project or www.census.gov) may have the capacity to store data in the terabytes. No matter how much data a given server can store, its purpose is to provide a means to share this information across the network. A server capable of storing huge amounts of data is useless if it cannot share the data with many users in a reasonable amount of time.

GIS data are often characterized by large file sizes and are shared through networks. It is frustrating for end users to know that the data is on the server, but they must wait several minutes for their retrieval and regeneration. The wait tolerance for each information product has already been specified in the IPD. When preparing to implement a GIS, consider your organization's requirement for both data capacity and data-transfer rates.

Data capacity is how much data can be stored; it is measured in megabytes or gigabytes. Network traffic is the amount of data traveling from the server to users; it is measured in megabits. When calculating the data volume moving from one storage disk to another over a network, the formula is: one megabyte equals eight megabits.

Data-transfer rate

Fifty-six Kb/sec is the same as 56,000 bits per second; so 10 megabits per second is the same as 10,000,000 bits per second. Thus, a 10 Mb/sec LAN can support transfer of nearly 180 times more data per second than a 56 Kb/sec WAN. This is often a misunderstood issue because of misinterpretation of communication units. But the devil is in these very details. Really understanding what the data-transfer rates mean is a critical dimension of GIS planning. Data-transfer speed is always the same (the speed of light), only the bandwidth, or throughput capacity, is different.

Client-server technology is a popular solution available to support network data transfer. The client requests data for an application and the server delivers the data to the application. The transfer of information relies upon a common language that permits communication between the sender and the receiver, called a communication protocol. A protocol is a set of rules that networked computers use to communicate with each other. The type of client-server architecture chosen for a system will dictate which communication protocol is called for.

Consider the following three types of client-server architectures:

Central server with workstation clients: A centralized file server shares data with computer workstations across the network. The application software resides on the workstations. Data is retrieved from the server and processed on the workstation. This type of system architecture requires the transfer of large amounts of data from the server to the client, creating high demand for bandwidth as a result. This type of configuration is best deployed over a LAN network environment. Common protocols associated with this network architecture: are NFS, SMB, CIFS, and TCP/IP.

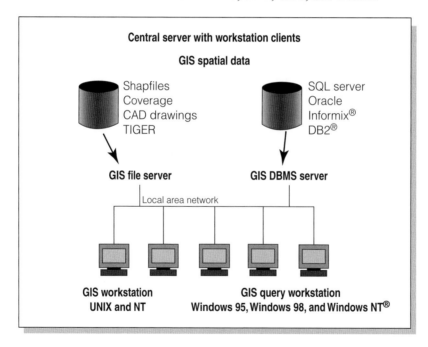

Centralized application processing with terminal clients: In this configuration, the data and application software are both stored and run on the server. The terminal workstations remotely control what application is performed. These are the "dumb" terminals that first brought computing to the people. The only data transferred from the server to the terminal is the resulting display data, which reduces bandwidth requirements. This type of architecture lends itself well to WANs where bandwidth is limited. Common protocols associated with this network architecture are RDP, ICA, X.11.

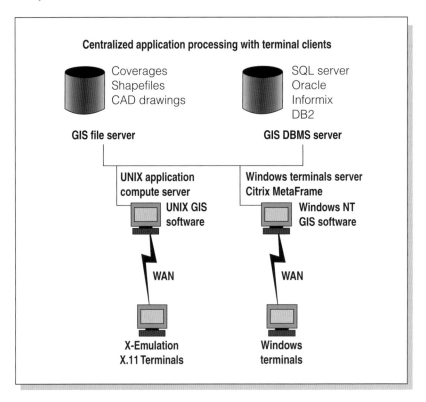

Web transaction processing: Application software and data files reside on a map server. The map server "serves" data and maps to Web browsers or other thin clients (e.g., Java™ applications) via the Internet or a secure local intranet. This architecture allows a single application process to provide simultaneous support to a large number of concurrent GIS users. Protocol associated with this network architecture: HTTP (Hypertext Transfer Protocol).

Each communication architecture has its own advantages and disadvantages. If your users require full transactional access to large data sets that they must check out, edit, and then check back, then a central server with workstation clients is necessary. If the information products must be generated quickly through a WAN, for example, to create a locator map for emergency response, then a centralized application with terminal clients is viable. If speed of data transfer is not critical, but the ability to share the information with a large volume of users is, then Web transaction processing is appropriate. Keep in mind that many organizations will use a combination of all three architectures to meet the demands of their user community.

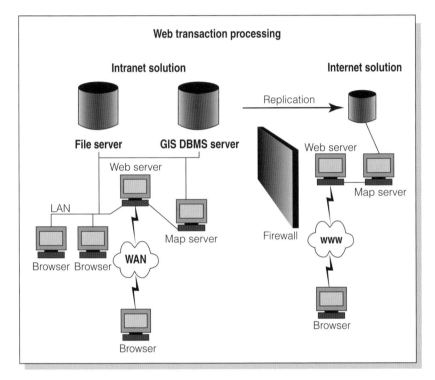

General issues of network performance

Networking is a specialized and quickly evolving field; some issues related to network performance are outside the scope of this book. The following discussion about networking capacities is intended to give you an overview of network traffic considerations, not a working knowledge of the intricacies of network engineering.

The volume of data transferred and network bandwidth capacity can be used to establish expected application response times to the user. A typical workstation GIS application might require up to 1 MB of spatial data that must be pulled over the network to generate a new map display or do an analysis. In contrast, a similar application deployed as a Web service or in a terminal environment might require just 100 KB of data to support the same display or analysis.

The following chart compares the network time needed for the data transfer under various configurations. It measures the total transfer of data in megabits (Mb) that must be transmitted and factors in data compression percentages that would typically be deployed under each client-server configuration. The transport time required to transfer these data are calculated for seven standard bandwidth solutions. This graphic illustrates how different client-server configurations affect the time it takes to transfer data.

Client-server communications		Network traffic transport time (seconds)						
		Wide-area network				Local network		
		56 Kbps	1.54 Mbps	6 Mbps	45 Mbps	10 Mbps	100 Mbps	1 Gbps
Data	File server to GIS desktop client (NFS)	893	32.1	8.33	1.11	5.00	0.50	0.05
	ArcSDE direct connect (DBMS client API)	18	0.6	0.17	0.02	0.10	0.01	0.00
	ArcSDE server to GIS desktop client (SDE API)	9	0.3	0.08	0.01	0.05	0.01	0.00
Display	Web server to GIS desktop client (HTTP)	18	0.6	0.17	0.02	0.10	0.01	0.001
	Web server to browser client (HTTP)	9	0.3	0.08	0.01	0.05	0.005	0.0005
	Windows terminal server to terminal client (ICA)	5	0.2	0.05	0.01	0.03	0.0028	0.0003

Best Practice

The first three client-server communication options in the chart are examples of a central server with workstation clients. In this case the full 1 MB of data is transferred from the server to the workstation to meet the demands of the application. The next two options represent Web transaction processing. The last option represents centralized application processing with terminal-client solutions. These last three cases make use of the data and application software on the server. While 1MB of data is needed to generate the information product the results are transferred from the server to the terminal clients as a display only, which reduces the amount of data required for transfer to 100 KB.

It is imperative to design a system that supports the total user base. The following table provides recommended design guidelines for network environments based on the expected client loads. These guidelines establish a baseline for configuring distributed LAN and WAN environments. Networks should always be configured with enough flexibility to provide special support to power users whose data transfer needs exceed typical GIS-user bandwidth requirements.

Local area networks Bandwidth	File server	SDE servers	Concurrent client loads X-Emulation	Windows terminals	Web products
10 Mbps LAN	5-10	10-20	50-75	150-300	150-300
16 Mbps LAN	8-16	15-30	80-120	250-500	250-500
100 Mbps LAN	50-100	100-200	500-750	1500-3000	1500-3000
1 Gbps LAN	500-1,000	1,000-2,000	5,000-7,500	1,000-30,000	1,000-30,000
Wide area networks Bandwidth	File server	SDE servers	Concurrent client loads X-Emulation	Windows terminals	Web products
56 Kbps Modem	NR	NR	NR	1-2	1-2
1.54 Mbps T-1	NR	1-2	9-12	25-50	25-50
6.16 Mbps T-2	3-6	6-12	30-45	100-200	100-200
45 Mbps T-3	20-45	30-60	250-350	700-1500	700-1500
155 Mbps ATM	75-150	150-300	850-1200	2500-5000	2500-5000
NR = Not recommended					

Determining your system interface and communication requirements

Now that you have a preliminary understanding of interface and communication technologies, you can think about the best configuration for your organization's system. Much of the information used for determining your system interface and communication requirements is located in the IPDs and MIDL.

Start by asking the following questions about your proposed system:

- What external databases, if any, are planned for use within the system and what format are they in?
- What records need to be accessed, how frequently, how quickly?

Consider any external databases that you plan to connect to your GIS applications. These data sets are identified in your MIDL, along with their storage format. The records needed, the frequency of access, and the wait tolerance are given in the IPD. Once these databases are identified and the access demands known, provisions can be made for obtaining the required interface software.

Next ask the following questions about your proposed system:

What are the wait tolerances of the information products?

Review the wait tolerances specified in the IPDs. Information products with low wait tolerances, such as emergency-service applications, will require a network solution that maximizes their production speed.

Where is the data located?

Identify the location of the databases planned for use with the GIS. The sources of this data are listed in the MIDL. In an ideal situation, all the data would be stored in one central database in a standard environment. In the real world, however, data sets will often be stored on different servers throughout the organization. Identify where the data is located, and if access to its location impedes the creation of your information products, find a solution.

Where are the data-handling locations and what is the data-handling load at these locations? At this point in the planning process you have already determined the location of all sites expected to make use of the GIS data. This issue was addressed while determining the scope of the system (chapter 7). You have also identified the number of users and the data-handling loads at these locations. This information is crucial for determining how to tie these users together through the network, and what communication infrastructure is required to handle the associated network traffic.

What is the current network configuration?

You should conduct a thorough survey of your organization's existing network environment, provided they have one. During this survey determine if the current network can support the requirements of the GIS, and if not, what enhancements are required to support GIS implementation.

What data volume has to be transmitted and when?

Using your information product display requirements from the IPDs, and the wait tolerance, estimate the total volume of data (Mb) that must be transmitted for your high-priority information products, particularly those with a low wait tolerance. Compare this demand with the existing bandwidth and traffic volumes in those parts of the network that will be used. Making these determinations allows you to estimate the interface and communication technology that meets the demands of your information products. The actual administration of the system, however, is best left to persons trained in this field. Your job is to identify the criteria required to make your system operate effectively. Provide these criteria to your network administrators and ask for their recommendations. If your organization does not have a network administrator, obtain the needed assistance from vendors or consultants. Even the most robust application and richest GIS data set will serve to frustrate users if network access becomes a bottleneck.

Distributed GIS and Web services

Distributed GIS is increasingly a dimension of new GIS installations. There are several steps in the thinking process that allow initial assumptions to be evaluated and a better understanding of the options and likely costs arrived at. The concepts that will be discussed are peak usage, peak bandwidth loads which leads to the assessment of bandwidth suitability, wait times, batch processing, platform sizing, and costs.

First review what is already known at this stage. For each department, for each business-process workflow, and information product, the degree of complexity of the information product and the number and size of data sets used has been determined. The same is true for the total number of users who will be making the information product, and of those the number that will be working concurrently. These can be brought together in a list (figure 2), which, for each site, summarizes the total number of users and the number of high-complexity and low-complexity concurrent users, and the peak Web usage expressed in number of requests per hour at each site.

To better show the use of this list, we've built an example for a typical city that we'll call Rome. In the first year it is initially proposed that the GIS will use the existing city communication structure. All GIS users will have to be linked to the central GIS DBMS database with search engine capabilities on the server. They will use the existing city system network configuration and bandwidth, some via Windows terminal servers and some through Web services. A site diagram of the year 1 configuration is shown on following page:

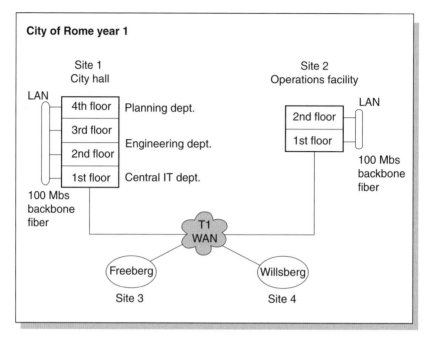

Figure 1   City of Rome existing user locations and network communications.
Central computing environment.

The number of high-complexity and low-complexity users that connect to the GIS city database and the Windows Terminal server platforms can now be identified. The total number of Web requests per hour can also be provided. The following table summarizes these figures (figure 3).

Year 2 requirements

The same estimates can also be made for the second year, perhaps assuming that new departments will need access to the GIS, or that there may be special requirements such as a firewall between the central computer and the police computer, or that operations will be extended to cover 911 emergency services and connections with vehicles, and that additional remote sites need to be serviced. Thus extended architecture options need to be considered for the second year.

City of Rome example—year 1

City of Rome—year 1				Total users	Peak usage		Web
					Processing complexity		Web
Department	Workflow	Info. product	User type		High	Low	Req/Hr
Site 1—City hall							
Planning	Zoning	1.0	Planner	20		8	
		1.1	Web service				2600
	Permits	1.2	Inspector	20		10	
		1.3	Appraiser	15	8		
		1.4	Supervisor	2		2	
		1.5	Web service				600
Engineering	Sewer backup	2.1	Engineer	4		3	
		2.2	Web service				800
	Elec. breaks	2.3	Electrician	13	6		
		2.4	Supervisor	2	1		
		2.5	Web service				600
	Hwy. repair	2.6	Field eng.	10		4	
		2.7	Contracts	4		4	
City hall totals				90	15	31	4600
Site 2—Operations facilities							
Operations	Clean-up prog.	3.1	Op. staff	4		2	
Operations totals				4		2	
Remote field offices							
Site 3 Freeberg	Inspection	4.1	Field engineer	40		30	
Site 4 Willsberg	Inspection	4.1	Field engineer	30		20	
Field offices	Inspection	4.2	Web service				1200
Remote totals				70		50	1200
City totals				**164**	**15**	**83**	**5800**

Figure 2   Year 1—peak usage by information product

Server platform	Peak Loads		
	High	Low	Web
City GIS database	15	83	5800
Windows terminal server	0	52	
Web services	Requests per hour		

Figure 3   Year 1—peak loads by server platform

Bandwidth suitability							
Network	Bandwidth	Estimated traffic (Mbps)			Peak loads		
Location	(Mbps)	Total	Desktop	Web	High	Low	Web req. per sec.
City hall							
City hall backbone	100.000	23.639	23.000	0.639	15	31	1.27
WAN connection (T-1)	**1.540**	**1.623**	1.456	0.167	0	52	0.33
Internet connection (T-1)	1.540	0.000	0.000	0.000			
Operations facility							
WAN connection (T-1)	1.540	0.056	0.056	0.000	0	2	0
Remote field offices (peak loads)							
Site 3 Freeberg	**1.540**	**0.923**	0.840	0.083	0	30	0.17
Site 4 Willsberg	1.540	0.643	0.560	0.083	0	20	0.17

Note change from Web req./hr. to Web req./sec.

Figure 4   Year 1—bandwidth suitablility using Peters network
design-planning factors

163

Year 2 configuration

The year 2 configuration maintains a central computing environment and uses an Internet solution for most of the additional services. A firewall is required between the central GIS data servers and application servers at the IT division in city hall and the police department. The database is replicated in the police department so that they may use their secure data in combination with their copy of the database to serve their office staff and, through wireless connections, their vehicles independently. A diagram of the year 2 configuration is seen below (figure 5).

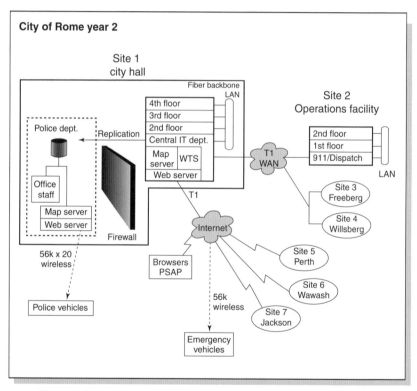

Figure 5   Year 2—proposed architecture (central computing environment, Internet solution)

Year 2 communications

The operations building is served from city hall over a T1 wide-area network (WAN) and it is proposed that the 911 emergency services be located there. The emergency services data are hosted on the city hall GIS data server and the emergency services desktop clients are provided terminal access to their low-complexity desktop applications running on the city hall terminal server farm, accessing data from the city hall data server. Emergency Web services are supported out of city hall on the enterprise Web server farm over the city Internet connection. Emergency vehicles access the 911 Web services over wireless Internet Web connections to the city hall Web services. Remote sites 3 and 4 remain connected to the T1 wide-area network, while sites 5, 6, and 7 are connected to the Web server at city hall.

The table on page 166 summarizes the increase in the numbers of users in both the high-complexity and low-complexity categories. Increases in the peak number of requests per hour over the Web can be compared from year 1 and year 2. Note that the loads on the GIS database server have more than doubled, the work being done by Windows terminal servers has tripled, and the number of Web server requests per hour has doubled.

Bandwidth suitablility

To assess whether the bandwidth available is adequate for the use proposed, a simple set of rules have been developed by D. Peters that link peak usage to peak traffic and hence bandwidth requirements (figure 9). These rules are based on the type of architecture involved.

The suitability of the bandwidths proposed now can be examined based on the peak loads anticipated and assumptions of the data per query and amount of compression to be used.

The six architectures covered are:
1. File server client—a standard GIS desktop client accessing data from a file server data source
2. DC client—a standard GIS desktop client with search engine capabilities connecting directly to a DBMS data source
3. AS client—a standard GIS desktop client accessing data from a server with a search engine on the DBMS server (used in the city of Rome)
4. Terminal client—access to GIS desktop software executed on a Windows terminal server (used in the city of Rome)

City of Rome example—year 2

City of Rome—year 2				Total users	Peak usage		
Site locations					Processing complexity		Web
Department	Workflow	Info. product	User type		High	Low	Req/Hr
Site 1—City hall							
Planning	Zoning	1.0	Planner	25	0	15	
		1.1	Web service				2600
	Permits	1.2	Inspector	25	0	15	
		1.3	Appraiser	20	10	0	
		1.4	Supervisor	5	2		
		1.5	Web service				900
Engineering	Sewer backup	2.1	Engineer	5		3	
		2.2	Web service				1000
	Elec. breaks	2.3	Electrician	13	6		
		2.4	Supervisor	2	1		
		2.5	Web service				1900
	Hwy. repair	2.6	Field eng.	11		7	
		2.7	Contracts	4		4	
City hall subtotals				110	19	44	6400
Police (Firewall)	Patrol sched.	5.1	Admin.	10		3	
		5.4	Web services				200
	Crime analysis	5.2	Detectives	10	5	0	
	Spec. events	5.3	Traffic	10		3	
Police totals				30	5	6	200
City hall totals				140	24	50	6600
Site 2—Operations facilities							
Operations	Clean-up prog.	3.1	Op. staff	4		2	
911	Response	3.2	Call takers	50		30	
		3.3	Web services				4000
Remote vehicles	Dispatch	3.4	Drivers	30		30	
Operations totals				84	0	62	4000
Remote field offices							
Site 3 Freeberg	Inspection	4.1	Field engineer	45		30	
Site 4 Willsberg	Inspection	4.1	Field engineer	60		40	
WAN field office	Inspection	4.2	Web services				1200
Site 5 Perth	Inspection	4.3	Field engineer	10		2	
Site 6 Wawash	Inspection	4.3	Field engineer	50		40	
Site 7 Jackson	Inspection	4.3	Field engineer	60		20	
Internet field offices	Inspection	4.2	Web service				1300
Remote totals				225	0	132	2500
City totals				**449**	**24**	**244**	**13100**

Figure 6   Year 2—peak usage by information product

Server platform	Peak loads		
	High	Low	Batch
City GIS database	19	238	20
Police GIS database	5	6	20
Windows terminal server	0	194	Web
City hall Web services	Requests per hour		12.900
Police Web services	Requests per hour		200

Figure 7   Year 2—peak loads by server platform

Bandwidth suitability							
Network	Bandwidth	Estimated peak traffic (Mbps)			Peak loads		
Location	(Mbps)	Total	Desktop	Web	High	Low	Web
City hall							
City hall Backbone	100.000	37.917	74.000	0.917	24	50	1.83
City WAN connection	6.000	3.863	3.696	0.167	0	132	0.33
Police WAN connection	0.056	56K dedicated dial-up connections			20 dial-up connections		
Internet connection (T-1)	1.540	0.921	1.736	0.736	0	62	1.47
Operations facility							
WAN connection (T-1)	1.540	0.896	0.896	0.000	0	32	
Remote field offices (peak loads)							
Site 3 Freeberg	6.000	0.923	0.840	0.083	0	30	0.17
Site 4 Willsberg	1.540	1.203	1.120	0.083	0	40	0.17
Site 5 Perth	1.540	0.070	0.056	0.014	0	2	0.028
Site 6 Wawash	1.540	1.203	1.120	0.083	0	40	0.17
Site 7 Jackson	1.540	0.643	0.560	0.083	0	20	0.17

Figure 8   Year 2—bandwidth suitability using Peters network design-planning factors

5. Browser client—browser client access to a standard Web image map service (used in the city of Rome)

6. Web GIS client—GIS desktop client access to a standard Web image map service

For each of these, assumptions are made (see appendix C) on the average data per query involved, including the percent compression applied. Converting data to traffic is expressed in kilobits per query (8 kilobits per KB and 2 kilobits overhead per traffic packet). The analysis assumes one user query every 10 seconds and results in average bandwidth utilization per user. Note that Web requests per hour are expressed as Web requests per second for ease of use with network load factors. These factors are summarized in the table on the following page.

It should be noted that the first three architectures in the table are normally supported on an ethernet local area network (LAN) environment. Maximum traffic on the shared ethernet segments is typically reached

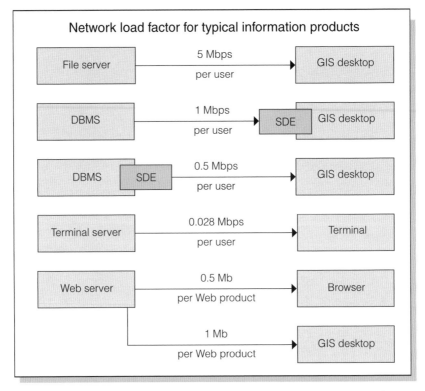

Figure 9   Year 2—bandwidth suitability using Peters network design-planning factors

around 20 percent to 30 percent of the total available bandwidth due to frequent collisions at higher utilization rates. Similarly, the second three architectures are normally supported over WAN environments. Optimum performance on shared WAN segments is typically reached around 30 percent to 50 percent of the total available bandwidth due to the probability delays of higher utilization rates.

Applying these considerations to the city of Rome allows the estimated peak traffic (Mbps) to be calculated and compared with the bandwidth available in the proposed network architecture.

Recommended year 1: It is clear that the WAN connection (T1) bandwidth is inadequate and that a T2 (6 Mbps) is recommended. Similarly, the T1 connection to the remote site at Freeberg will experience delays at higher utilization rates (optimum use is below 30 percent to 50 percent of total capacity) and an upgrade to a T2 is recommended on that link.

Recommended year 2: There is a considerable increase in peak traffic, putting pressure on both the city hall backbone and the terminal server connections when percent utilization delay factors are taken into account.

Assuming that the year 1 upgrades have been made, in year 2 it is recommended that the city hall backbone be upgraded to a gigabit switch. This switch will not be expensive as the existing backbone is on fiber.

The WAN connection, already a T2, will be under stress with 132 users and consideration should be given to upgrading it to a T3 (45 Mbps).

The WAN connection from the operations facility and the new Internet connnection from city hall should be established on a T2 (6 Mbps) to accommodate the peak traffic, as should the connections to the remote sites at Willsberg and Wawash.

These recommendations will have a significant impact on the costs of GIS communication in the city of Rome, and must be taken into account in the cost model.

Wait tolerance assessment

Each of the information products has a wait tolerance defined in the information product description. It is now possible to review the architecture and assess whether it will allow the wait tolerances to be met. This is a balancing act. Clearly the time taken for disk access, application CPU processing and video display will determine wait tolerance on a stand-alone workstation. In distributed processing disk access, the server CPU processing, the network communications, the application CPU processing and the video display will be involved. The total response time of a particular application query will be a collection of the responses from each of these components. With current CPU technology, most application processing and server processing takes less than two seconds. In the case of distribution processing, however, the data transport time over the network may be the primary determinant of network transaction performance.

Client-server performance numbers again provide the network traffic transport time for each bandwidth in six combinations of server-to-client transmission.

The backbone in city hall, where file servers support applications being run on Windows clients, has a response time of 0.5 seconds. The Windows terminal servers to terminal clients over a T1 WAN, have a response time of 0.2 seconds, or if a T2 WAN was installed a 0.05 second response, and the Web server to browser clients in PSAP units and remote sites 5, 6, and 7, have a response time of 0.013 seconds. Web server to emergency vehicles and police vehicles over 56 k wireless transmission have a response time of 9 seconds. The accumulation of these times can be calculated for individual information products.

Other software applications will impact this bandwidth and must also be taken into account.

Batch processing

Any system of the type envisaged carries out some functions in a batch processing mode. These include data loading, database administration, data backup, replication services (for the police), reconcile and post operations for the engineering department, and some automated map production. These will be done in the IT department in the Rome city hall, but will make demands on the CPU. While this can be scheduled in off-peak hours, the best practice is to assume that one batch process requires a dedicated CPU on the server for the time during which it is running. Batch processes executed on a separate client (workstation, etc.) generally require much less than one server CPU (up to five concurrent client batch processes can be supported by one single server CPU).

Platform sizing

This leads to the assessment of the number of CPUs required for the whole system and the platform sizes required. To help in this process another set of useful rules is provided in the CPU performance model in figure 10. In compiling performance models, it is understood that when a typical desktop client accesses a data server, for example, a certain percentage of the processing is carried out on the desktop and a certain percentage is carried out on the server. This results in calculations of how many desktop clients can be supported simultaneously by one server CPU. Clearly, the nature of the desktop and the software being used impact this distribution of effort and, for the performance models given, all CPUs are assumed to be the same (2400 Mhz). These will scale in a linear fashion.

Platform	Performance per CPU	Memory
GIS file server with search engine	20 Clients/CPU	1 GB/CPU
Web transaction loads on the data server	1200 transactions per hour equivalent to 1 data server client	
Internet Web server	File server data source 3,000 transactions per hr./CPU	512 MB/CPU
	Data source with search engine 6,000 transactions per hr./CPU	
Windows terminal server	File server data source 4 GIS desktop users/CPU	1 GB/CPU
	Data source with search engine 8 GIS desktop users/CPU	2 GB/CPU

Figure 10   Performance per CPU 2003

The summary of the number of CPUs required and the platform sizes is given in figure 11. In year 1, the city GIS database has peak loads of 15 high-complexity users and 83 low-complexity users. Based on the CPU performance models, it is known that a single data server CPU can support 20 concurrent desktop users.

Web transaction loads on the same data server can be included with standard desktop clients through a simple conversion (one desktop client equivalent to 1,200 Web requests per hour).

Batch processing would also be done on this computer. There are four batch processes envisaged in the first year: database loading, database administration, data backup and reconciliation, and post operations for the engineering department. If an eight-second response time was considered to be adequate, a single CPU can handle these batch requirements. Only one batch process would be allowed on the server during peak loads to support client performance requirements. Web loads similarly make an impact on the CPU. This results in a total load of 123 GIS-related users on the city GIS database in year 1, requiring 6.15 CPUs.

Terminal server performance is estimated in the same manner. In the first year there were 52 low-complexity users. Using the performance model rule of eight per CPU, 6.5 CPUs are required.

Internet Web services are estimated as having a peak requests per hour of 5,800 in the first year. With the performance rules of 6,000 requests per hour per map server CPU accessing a DBMS data source with a search engine, the demand could be met by 0.97 CPUs.

171

Rome—year 1 and 2. Server requirements

Server requirements and costs

Year 1

Data server capacity	Peak desktop		a Batch load	b Web load	Total load	c Per CPU factors	d CPU sizing estimate	e Estimated pricing	
	High	Low						Intel	UNIX
City GIS database	15	83	20	5	123	20	6.15	$80,000	$100,000
City Web services			Requests per hour		5,800	6,000	0.97	$12,000	$12,000
WTS farm	0	52			52	8	6.5	$48,000	NA

Year 2

Data server capacity	Peak desktop		a Batch load	b Web load	Total load	c Per CPU factors	d CPU sizing estimate	e Estimated pricing	
	High	Low						Intel	UNIX
City GIS database	19	238	20	11	288	20	14.39	$240,000	$360,000
City Web services			Requests per hour		12,900	6,000	2.2	$24,000	$24,000
Police GIS database	5	6	20	0	31	20	1.6	$12,000	$12,000
Police Web services			Requests per hour		200	6,000	0.03	$12,000	$12,000
WTS farm	0	194			194	8	24.25	$144,000	NA

Figure 11   Storage pricing estimates: volume of data (Gigs) + 50% (database indexing) X $ per gig at RAID level required.

a. Batch load: batch process in replication takes full CPU = 20 clients

b. Web load: divide req. per hour by factor of 1200 = number of clients: 5800/1200=5

c. Per CPU factors: number of clients/requests per CPU

d. CPU sizing estimate: Use hardware pricing model 2003 provided

 Note that SRint 2000 specification indicates current CPUs

 Choose next highest CPU number for intermediate values, e.g., 6.5—use 8

 WTS farm CPUs can be used most efficiently in units of 2 CPU, e.g., 6.5=2x4 @ $12,000=$48,000

Hardware pricing model: 2003			
Performance		Operating system	
CPU	SRint2000	Intel	UNIX
2	18	12,000	12,000
4	18	30,000	30,000
6	34	55,000	68,750
8	42	80,000	100,000
10	50	120,000	150,000
12	57	160,000	200,000
14	63	200,000	300,000
16	69	240,000	360,000
18	76	280,000	420,000
20	85	320,000	480,000
22	94	360,000	540,000
24	103	400,000	600,000
26	112	450,000	900,000
28	131	500,000	1,000,000
30	140	550,000	1,100,000
32	149	600,000	1,200,000

Figure 12

A different picture presents itself in year 2. Use of the city GIS data server now totals 257 concurrent users. The batch-processing loads include a greater level of activity with the addition of replication services for the police department and automated map production. While these can be fit into the business operations schedule and can be scheduled at off-peak times, they still make significant CPU demands. A single batch process on the server consumes a server CPU, which is equivalent to the server load supporting 20 concurrent desktop clients. Web loads have increased to 18 equivalent users for a total load of 295 users on the central data server requiring 14.7 CPUs. The police department has a separate secure database and services six office staff users and batch processing including replication, for a total of 31 equivalent users requiring 1.6 CPUs.

Terminal server loads are significantly extended in year 2. There are additional servers in city hall and in the police department, for a total of 194 concurrent users. At eight per CPU, this will require 24.3 CPUs.

A total of 12,900 peak Web requests per hour are estimated in year 2; 6,000 per hour can be accommodated per CPU accessing the city hall database management system (search engine) database, leading to a demand for 2.2 CPUs. The police department provides its own secure Web services to its vehicles over 56 k dedicated dial-up lines. This is a modest use but requires 0.03 CPU.

Costs

These server requirements and related costs (in 2003) are summarized in figure 11 (see page 172). The costs for servers are taken from the hardware pricing model (figure 12). The type and number of desktop and browser units can be estimated from the total number of users in each level of processing complexity from the analysis completed above (see the prices given on page 92).

Software license costs for all systems, particularly distributed GIS architectures, are dependent on the level of complexity of the software times the number of concurrent users. These calculations can be made from the analysis completed above.

Storage costs are dependent on the volume of digital data taken from the MIDL plus 50 percent to allow for data indexing at a price per gigabyte dependent on the level of security required (see page 96).

It should be recognized that these are preliminary estimates to provide platform sizes and communications bandwidths to allow first approximations of cost to be used in the benefit-cost analysis to follow. It must also be noted that the engineering and pricing models change over time as technology changes and more benchmark test results and user experience become available. The design process is a balancing act between workstation performance and its related software on one hand, and server performance, database design, and storage on the other hand, with the varying demands that different configurations would make on network communications bandwidth. The optimizing of the architecture and platform and bandwidth usage involves optimizing the workflows, the database configuration, the versioning methodology, the database design, editing operations, and having the ability to identify problems, correct performance statistics, and monitor table statistics. A good database administrator is essential to this optimization.

It is clear from the straightforward comparison of the technology needs in year 1 and year 2 of the example of the city of Rome that there can be a phased implementation strategy for the acquisition of technology and its deployment year by year. No longer is the paradigm of buying one large computer and expecting it to handle all requirements for the next five years a useful paradigm. Instead there must be a strategic plan for managing technology change integrated with the activity plan for data acquisition and application development provided earlier, and being fully cognizant of the technology lifecycles that can be anticipated.

Hardware and software

At this stage in your planning you should have a thorough assessment of your hardware and software requirements. In chapter 7, you determined your workstation requirements and storage requirements. In this chapter you established your software function requirements and your interface and communication needs. What remains is to consider these requirements in the context of the existing policies or standards that your organization has adopted and to examine your hardware and software procurement requirements, taking into account technology lifecycles.

Organization policies and standards

It is always important to comply if possible with any policies and standards your organization may have adopted regarding system configuration. For example, it is important for you to know if your organization has adopted an operating system standard (e.g., Windows 2000, UNIX) or if they have a specific processing model (e.g., Web processing). Document these preferences in the preliminary design report and the request for proposals.

You will usually be required to design your hardware and software and communication needs in the context of existing policies and practices. If you depart from these, the case for additional benefit must be strong.

Many organizations have a long-established relationship with particular hardware or software vendors. These relationships have developed because of the past reliability and suitability of the products and the level of service satisfaction and trust in particular vendors. Your organization may already have established maintenance agreements (common in big enterprise hardware and software deals) or may have invested heavily in training for certain products. These issues cannot and should not be ignored. Do, however, always bear in mind that normal technology lifecycles dictate that no adopted solution should ever be considered permanently "set in stone."

If change in technology platforms or operating systems are required, the reasons for these changes must clearly be spelled out. The critical inadequacy of existing capabilities must be explained. It is not uncommon for the specter of an enterprise-wide GIS implementation to invoke major changes in organization-wide technology standards.

Technology lifecycles

The lifecycle of technology, including the rate of technologic change, is a crucial consideration in the acquisition, purchase, or upgrade of new systems. Computer technologies are constantly evolving and becoming more cost-effective (we've all heard of Moore's Law by now). Keep senior management aware of this situation and your proposed strategy for keeping the GIS cost-effective. If management knows well in advance that future upgrades will be required, they will be that much more likely to approve funding for regular technology upgrades.

So what is the actual lifecycle of various typical technologies? The following table (prepared in 2003) shows the technology lifecycles for all of the technologies related to a major GIS implementation. For each major technology, it shows a time period in months for which that technology would be considered current, useful, obsolete, and nonfunctional. Current is the period between major releases with significant functional improvement; useful is the length of time that current software will run on this equipment; obsolete indicates when new releases of software will not be compatible with the equipment; and nonfunctional is defined as the point at which the technology is no longer functional, the cost of maintenance is greater than the residual value of the equipment, and training is a waste of money. A careful study of the figures reveals that networking technology and operating systems have the longest lifecycles (up to three years of currency). Contrast this to workstation, desktop,

2002–2003 Estimates of technology lifecycles (months)				
Technology	Current	Useful	Obsolete	Nonfunctional
Network infrastructure	24-36	37-84	85-120	120+
Wide-area networks*	12-24	25-60	61-84	84+
Computer				
• Server	12-18	19-48	49-72	72+
• Workstation	6-12	13-48	49-72	72+
• Desktop	6-12	13-36	37-60	60+
• Laptop or mobile PDA	6-12	13-24	25-48	48+
Operating system software	18-36	37-60	61-72	72+
Vendor software	12-18	19-36	37-60	60+
Internet products (browsers, associated products)	9-12	13-24	25-36	36+
Data	Variable—depends on rate of decay of validity			
*Internet bandwidth increasing at 300% per year				

laptop computers, and mobile PDAs which can be expected to move from current to useful in much shorter periods.

To stay current with technology trends, attend software user conferences, read current GIS publications, and establish communications with vendors and industry peers.

The preliminary design document

At this point in your GIS planning you have finally determined the data and technology requirements and you are now ready to place your findings into a report. This report covers your complete work so far and will mark the transition from system design to system procurement and implementation. The preliminary, or conceptual, design report spells out the requirements for the GIS that must be put in place to meet your organization's needs. This preliminary design document is a critical component of your effort.

The conceptual system design is derived from the function requirements identified in the information product descriptions (IPDs), the data input requirements as identified on the master data input list, and the conceptual system design for technology (addressed in the first section of this chapter).

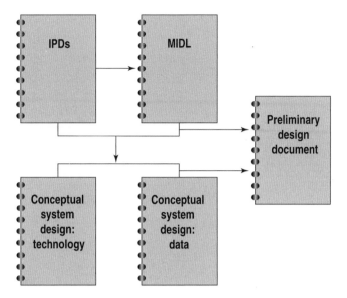

The planning process builds upon itself, with the results from earlier work carried through for use in later stages.

Include the following sections within the preliminary design document:
Executive summary: Begin the report with an executive summary. This portion of the document summarizes the findings and recommendations of the report. Senior officials, who are short on time and low on GIS knowledge, often will only read this section of the report.

Although the executive summary appears at the beginning of the preliminary design document, it is not created until after the completion of the entire report. Once the report is completed, review it in its entirety, paying close attention to the recommendations section. Gather this information into summary form and create the executive summary.

Introduction: Immediately following the executive summary, include a brief introduction to the report. This section should quickly describe the report's purpose and structure. Include a table of contents within the introduction.

Data section

The structure detailing the design and development of the database and its impact on system design. The data section of the preliminary design document contains the following subsections:

Data set names

There is little to build on without first identifying the data sets. Document the name of each data set. Obtain this information from the MIDL.

Data characteristics

Within the report identify the physical and spatial characteristics of each data set. Report the medium of the source data (e.g., paper maps, Mylar sheets, CDs, tape disks, or floppy disks). Identify the digital data format, if the data is already in digital form, and report its format (e.g., TIGER, .E00, .DXF, .txt). Report the size of each data set, map projection, scale, and datum. Pay careful attention to note any data sets that will require conversion to the chosen projection or scale. Finally, be sure to report the type and amount of error tolerance for each data set. Gather all of this information from the MIDL.

Logical data model

The conceptual database design is a high-level view of how the database will work. The logical design is a preliminary layout that fills in the conceptual design in accordance with a specific data model. For the conceptual database design report the database contents, including all of your data elements and their logical linkages. You can also begin organizing the data thematically, such as land ownership, transportation, and environmental areas. Include diagrams of the relationships between data elements. This step will give you an idea of the overall database and the complexity of its data relationships.

Conceptual system design for technology

The technology section of the preliminary design document defines the remaining components of the conceptual design for your organization.

This section clearly identifies your planned utilization of the technology and contains the following subsections:

Function utilization

Report the function utilization in the main document or as an appendix. It should contain an overview of the software functions required to create the information products needed. Gather this information from your conceptual system design for technology. It is useful to include the list of function requirements.

System interface requirements

Describe the system interface requirements identified during the conceptual system design for technology. For example, your GIS applications may require a link to scanned images and documents managed by an existing document management software program. The need to access these databases must be documented. Include detailed information about the specific software as an appendix if needed.

Communication requirements

This subsection includes a review of the existing communication infrastructure and identifies the network communications required to support the GIS. Also report all locations within the organization requiring access to GIS and the number of GIS users at each location. This information was first established in chapter 7 (Define the system scope) and further discussed in the network communications section of the system design chapter 10. Return to this work now and place the findings within the report.

Hardware and software requirements

Report the existing hardware and software standards and policies of your organization as well as the proposed hardware and software configuration for the new GIS. If required, discuss any system configuration alternatives based on your organization's policies and standards.

Document any compatibility issues with existing computing standards and system configuration that have been identified. It is useful to always consider the integration of this proposed configuration with other systems used, or planned for use, within the organization.

Policies and standards

Report the policies or standards regarding the implementation of technology that exist within your organization. If your organization does not have any specific policies or standards affecting the conceptual design, note this in the preliminary design document.

Recommendations

The last section of the preliminary design document (we can think of the appendixes as a supplement to the report) is the clear-cut recommendation for a system design. All the previous sections of the preliminary design report were a straight reporting of findings from your earlier work. This section is where you, and your supporting team, provide actual recommendations for the system's final configuration.

Base your recommendations on logical decisions. Draw from the data and technology section of the report, use your staff to help with the decisions, and make the design recommendations for both software and hardware technology. You are recommending "what" is needed to meet the needs of your organization. At this point you are recommending the overall design required, not a specific choice among the offerings of vendors and suppliers. In fact, avoid that as much as possible at this stage.

Make the recommendations clear and concise. You have reached a critical point in the planning process. The assessment of needs, data, and technology is completed. The recommendations take all of this into account and provide a basis for asking approval to proceed with the procurement planning and implementation planning.

Appendixes

Anything that is not needed to understand the core of the design is a candidate for an appendix. For example, it's generally good to provide all the requested IPDs and the MIDL as appendixes.

Include as many illustrations and diagrams in the document as are needed to make the points. For example, visual representations of map layers help people understand the nature of the data selections, while communication networks in the technology section are most clearly understood in the form of schematic diagrams.

Once a draft of the preliminary design document is complete, circulate it for review and comments to the individuals involved in the GIS database design. When you are implementing a shared database model, representatives from each department planning to use the system must review your plan. This review is performed to ensure the overall consistency and completeness of the database concepts.

After this review and comment period, the preliminary design document must next receive formal approval from the GIS committee and senior management. When this stage of the conceptual system design is complete, there will still be a significant number of details to consider. But before you can move into the actual procurement and implementation stages, there must be general agreement on the overall system design.

This chapter described the purpose and contents of the preliminary design report. The report is based upon the previous work completed during your planning phases. It pulls all your previous planning work together and provides recommendations for the system's overall design. Completion and approval of the preliminary design report permits the transition to the procurement and implementation phases of GIS planning.

Benefit-cost, migration, and risk analysis

"The most critical aspect of doing a realistic benefit-cost analysis is the commitment to include all the costs that will be involved. Too often managers gloss over the real costs, only to regret it later."

Benefit-cost, migration, and risk analysis

Benefit-cost analysis is a technique that allows you to compare the expected cost of implementing a system with the expected benefits that will result from having the new information products. The result of this comparison will be an indication of whether or not your project will be financially viable, and when you can expect an actual financial return on your initial investment.

There are four stages in benefit-cost analysis, each of which is examined in this chapter:

- Identifying costs by year
- Calculating benefits by year
- Comparing costs and benefits
- Calculating benefit-cost ratios

Identifying costs by year

Benefit-cost analysis happens only in the context of a specific time period. The cost model documents all the costs you expect to incur during your GIS implementation for a specific planning period. The thing that makes it work is a commitment to include all the costs that will be involved. The cost model breaks all costs down to one of five categories:

1. Hardware and software

Estimate the cost of your yearly hardware and software requirements. Include workstations, servers, disk drives, CD-ROMs, tape backups, upgrades to existing hardware, input devices such as digitizers and scanners, output devices such as laser printers and plotters, and software licenses. Maintenance costs for your hardware and software are also a significant component and should be included in the cost model. The scheduling of incremental technology acquisition is the key to subsequent financial management.

2. Data

Depending on the project purpose and scope, data acquisition can easily become one of the most significant components of your system and consume a high proportion of your first five years' funding. The costs are associated with data acquisition, editing, conversion, updates, and maintenance. From your requirements study you already know what data you need to acquire and the costs involved.

3. Staffing and training

Do you have adequate staff to support the project? If not, you need to estimate the cost of hiring and training additional staff as well as retraining existing staff. Make sure to include the costs of travel associated with training. Over time the cost of staff will be your single largest financial outlay.

4. Application programming

Consider the cost of writing application programs, even simple ones. Do you have the right staff to write application programs? If not, you may have to hire a vendor to create what you need.

5. Interfaces and communications

Finally, calculate the cost of any hardware or software interfaces you need to purchase and any communications networks that are necessary to support your system.

Surprisingly to the uninitiated, the immediate initial hardware and software costs typically comprise a relatively small percentage of the overall cost model. Staffing, data, and application programming all require significant up-front expenditures. Figure your costs for at least the first year, and better still for 1-, 2-, 3-, 4-, and 5-year windows to see what will happen over time.

Now, the value of the planning process starts to show (or its absence felt). Since you have described the information products that are needed and have planned the data readiness that they depend on, you can identify things like the timing and size of the application program development budget needed with some confidence. There have been well-publicized examples in recent years of major GIS technology procurements that failed to specify information products up front so they had not factored in application programming. Costly delays and other unanticipated expenses resulted. The worst part is that the poor planning doomed the efforts before the very needed new business processes (manifested through the GIS applications) could be deployed.

To create a cost model for your project, set up a cost matrix by year or time period that breaks down the five cost categories by year that the cost will be incurred. Establish the columns for each year, list the data that will be available in the first year, and determine the associated hardware and software requirements that will be needed to produce the information products planned by year. As your database grows, more information products will be generated. To keep pace with the growth of your system, look ahead at expected demand and plan accordingly. Avoid buying any hardware that you cannot expect to use at least 50 percent of its capacity in the first year.

Calculating benefits by year

Enough empirical data now exists such that benefit analysis in GIS can be relied upon as an economically rigorous reality check in the planning process.

Recall the major categories of benefits:

- Savings: savings in money currently budgeted (i.e., in the current fiscal year) through the use of the new information provided by the proposed GIS. (e.g., reductions in current staff time, increases in revenue)
- Benefits to the organization: these benefits include improvements in operational efficiency, improvements in workflow, reduction of liability, increases in the effectiveness of planned expenditure, and increases in revenue. This is perhaps the most significant category of benefits in the worldwide use of GIS
- Future and external benefits: these are benefits that accrue to organizations other than those acquiring the GIS. To date, these benefits have often been underestimated in benefit-cost calculations

You have already estimated the benefit that will result from having each information product. The plan developed provides the estimated time of availability of each information product.

The question to be answered now is when will the benefits from each information product accrue to the organization? This will depend to some extent on the nature of the benefit, but as a conservative rule of thumb it is recommended that the benefits be calculated as being available one year after the first availability of the information product concerned.

Comparing benefits and costs

After you have estimated the costs of system implementation and the benefits that will accrue from having the information products created by the system, you can assess the benefits in relation to their associated costs. This assessment is the crux of the benefit-cost analysis. It will tell you if your implementation is financially feasible and, if so, when you will see a real return on your GIS investment.

You can use graphs like the ones below to visualize costs versus benefits over the life of the project.

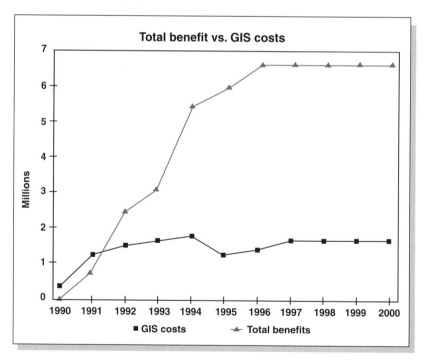

In this first graph, all dollar values are presented in nondiscounted dollars. Costs are represented by the black line which peaks early at just under $2 million, then falls gradually over the period under study. Meanwhile, the benefits shown by the gray line, rise steadily through the early years then level off at a high value over time. In this example, the graph shows that implementing a GIS requires a high front-end investment mainly during the first years, with a positive net effect approximately two years into the project.

This type of data can also be examined as a pair of cumulative numbers. The chart below displays cumulative costs and benefits. The new benefit or cost incurred each year is added to the previous year's total to give the new figure. All figures have also been discounted (brought back) to 1991 equivalents to remove the effects of inflation and reflect the true net present value of money. The cumulative benefits first exceed the cumulative costs in 1995. The overall benefits of a well-planned and well-managed system typically begin to outweigh the total costs somewhere between the fourth and the sixth year.

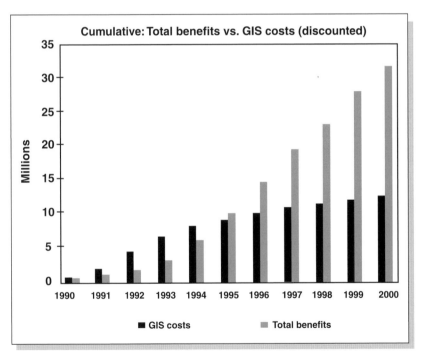

The cost model and benefit-cost analysis should be drafted on a department-by-department basis to allow alignment of departmental budgeting plans with the acquisition of the GIS.

Calculating benefit-cost ratios

Express all costs identified in your cost model in discounted funds. Costs that spread over a period of years into the future must be discounted to the base year using a discount rate to remove the effects of inflation and capture the true value of money. For example, suppose the base year is 2000 and the cost of a maintenance contract is $100 per year.

Year	2000 (base year)	2001	2002	2003
Actual cost of maintenance contract	$100	$100	$100	$100
Cost discounted to base year (using discounting rate of 7%)	$100	$93	$86	$80

The discounted costs, shown in the bottom row, are used in the benefit-cost analysis.

The base year is the current year or the year in which the benefit-cost analysis begins, whichever is the earliest. Because of inflation, if $100 of benefit accrues in the year 2005, this is worth less in real terms than a benefit of $100 realized in 2000. Similarly, a cost of $100 incurred in the year 2005 costs less in real terms than a cost of $100 incurred in 2000. To assess investment decisions properly, both the costs and benefits must be discounted to a common year—the base year—to remove the effects of inflation.

A discount rate of 7 percent is often used as a fairly standard average. Thus, $100 last year is worth $93 this year. Check the discount rate being used in your organization. Identify the benefits occurring over time and give these values discounted to the base year, as you did for costs. Calculate the net present value and benefit-to-cost ratio for each year of your analysis. The net present value (NPV) is calculated by subtracting the present value of costs (PVC) from the present value of benefits (PVB).

$$NPV = PVB - PVC$$

A positive NPV is the usual decision criteria applied for acceptable projects. The benefit-to-cost (benefit:cost) ratio is a fraction that represents the return (or benefit) on the investment (cost). Usually, the denominator (the cost) is reduced to 1.

$$B:C = PVB/PVC$$

For example if a $100,000 investment yields a return of $260,000 the benefit-to-cost ratio would be **260,000:100,000** or **2.6:1**. Investments that have a ratio greater than 1:1 will at least break even.

You can assess how changes in key parameters might affect the over-all results. To do this, you need to rework the calculations for different implementation scenarios. These scenarios might incur different costs or lead to different benefits that must be accounted for in the analysis. This is called "sensitivity analysis."

These basic steps have been adapted from those presented in Smith and Tomlinson (1992). For more detailed information on their published methods conducting GIS benefit-cost analysis, refer to the *International Journal of Geographical Information Systems* 6(3): 247-256.

Migration strategy

Crafting a good migration strategy is highly dependent on the scope of the effort. Obviously the installation of a small system in one department that has never had a GIS requires a radically different strategy than a major "roll-out" of an enterprise-wide system in a large organization. In almost all cases, however, the new GIS will be implemented in the context of existing data-handling systems that need to migrate from the old to the new. So regardless of the scope, you need to plan the sequence of events in this process and provide timelines for dealing with the old systems as well as the new.

Legacy systems and models

Legacy systems are an IT euphemism for an existing (sometimes quite old) technology platform that will be phased out by new technology. But legacy systems also represent the current ways of doing things, and these things represented by the business processes used to carry them out must also be migrated. Complicating matters, because they support ongoing operations, these systems and processes must be migrated seamlessly into the new system with a minimum disruption to business. This is quite a tight rope to walk.

Chapter eleven

Just as you have allowed time in your planning for new system acquisition, training and break-in, so must you allow time in the legacy systems for phase-out and reallocation of resources. This usually means a planned period of overlap between legacy systems and the new system. The annals of IT professionals are littered with the stories of organizations which terminated critical legacy systems prematurely on day one of new system implementation with disastrous results. The key to not joining their ranks is not to pull the plug on any legacy system until it can be proven that all of the needed data and functionality has been reliably migrated to the new platform.

The migration of existing business models can cause a real challenge. Many existing business models and processes have taken several years to develop and refine. Proven science is built on a set of functions and operations that can be carried out using existing systems and techniques in the organization. Since the need for the business model or process continues, there are three options for migration:

1. Rebuild the model so that it can be implemented in its present form into the new GIS-centric technology.
2. Improve the model, perhaps by replacing some of the assumptions with actual measurements using the GIS.
3. Abandon the model to design and build an entirely new model that takes advantage of GIS capability and makes possible new workflows.

Some of the questions that you will ask yourself are:

- How difficult is it to map the process or model into the GIS data structures?
- How long would it take to do this?
- Who will do it? Are the people who designed the original model still available?
- What are the new GIS capabilities that you would like to access—can they be integrated into the old model, or is a model redesign and rebuild necessary?

Once you have considered these issues, an appropriate migration strategy will emerge.

The migration from legacy systems to new GIS architectures is extremely dependent on the vendor offerings, both those already in place and the new ones to be acquired. The legacy system of a file server supporting the high-complexity ArcInfo workstations and low-complexity ArcView workstations needs to be migrated to a geodatabase with new data search engines, new object-relational data models, and edit versions for reconcile and post or versioning operations. ArcGIS workstations using their own personal geodatabase on a file server and browsers interfacing with ArcIMS® providing Internet services are parts of the new architecture. The business-process data models can be accommodated in the new object-relational database. Avenue™ macros previously used at the ArcView level need to be rewritten in Visual Basic to work in the new environment.

Migration considerations

When moving from old to new systems or applications, present management with a number of new considerations:

- Age: is the system or application slowing the forward progress of organization?
- Cost: cost of a new system; cost of developing a new application; cost of transition; people availability.
- Benefits: is it necessary to do this transition now? Benefits to be gained by the transition.
- Future transitions: will the proposed transition meet future business needs? Will legacy applications work alongside new applications developed on new GIS in the future?

There are also technical considerations. It is easy to find oneself extremely dependent on vendor offerings, both those in place and the new ones acquired. The movement to enterprise operations also involves dealing with data standards and data integration issues related to mapping from one schema to another. Gap analysis is the definition of any functions required by any information product that are not yet available in the new GIS. Expectations must be managed as there will be waits required for custom applications to be built.

Pilot projects

Some organizations are more comfortable taking the GIS plan and implementing it incrementally via pilot projects. A pilot project is essentially a test run of a small-scale version of your planned GIS. For example, an organization with headquarters and offices in multiple regions may want to establish the pilot GIS in headquarters and in one of the regions to build experience and gain an understanding of the administrative and communications problems that may be encountered before wider implementation. Similarly, within one department it may be wise to focus on a selected subset of information products or even subsets of the database as first steps in implementation.

Pilot projects are useful to demonstrate the planned GIS to management and potential users. They also can serve to evaluate the performance of the intended system, solve data problems before the final cost model is developed, and verify costs and benefits.

Whatever its purpose, a pilot project should have clear objectives. Often they are not conducted for the right reasons. Vendors tend to encourage pilot projects because once your organization brings in a particular system, it is seen as more unlikely that you'll implement another system. In this way, pilot projects may strongly bias the procurement process. Senior management likes pilot projects because they think they can show that they are tech savvy without spending too much money. Be wary of the ploy of using a pilot project as a way to get GIS proponents off their backs. Lower management tends to like pilot projects because they can get the GIS moving without the "hassle" of a thorough full-planning process. Clearly stating the objectives will help you avoid these pitfalls of the piloting process.

Avoid pilot projects if any of these are the motivating factors. An appropriate pilot project should always be launched in the context of your planning, and should in fact be part of the plan from the beginning. If you have adequately planned and now want to start implementing incrementally, a pilot project may be the perfect solution.

Risk analysis

To ensure a successful GIS implementation, it is important to consider and assess the risks associated with your implementation strategy. Risk assessment involves a thorough evaluation of your implementation strategy. You must understand the risks associated with your implementation strategy and the potential for project failure. The basic approach to risk analysis is:

1. Identify the type of risk involved in your project.
2. Discuss the nature of the risk in the context of the planned implementation.

3. Describe the mitigating factors that will be used to minimize the risk.
4. Assess the likelihood and seriousness of each risk to the project and give it a score.
5. Summarize the score—evaluate overall acceptability of the risk of the project.

Step 1: Risk identification

When seeking to identify the risks involved with your project evaluate the following factors:

Technology

- Is the technology being adopted new? The first release of software or hardware can sometimes have bugs or flaws.
- Are there gaps in the technology that prevent it from fully supporting your needs?

If there are gaps in the technology, do you need to enter into a contract with the vendor before acquiring the system to ensure that the gaps are filled? If the proven technology cannot create more than 80 percent of your information products, you are in a high-risk category.

Organizational functions

- Are there any functional changes in the mission of the departments or changes in departmental functions that can be foreseen?

These changes can take a long time and add complexity to your project. Increases in complexity will increase risk.

Organizational interactions

- Organizationally, are multiple agencies involved?
- Are they geographically dispersed?

Working with multiple agencies or multiple locations adds complexity to your implementation.

- Are changes in management required?

These changes may take a long time; you need to determine if you can succeed without change.

Constraints

- Are there budget constraints?
- What is the timing of the project?

Adequate funding and a realistic timeframe are essential for success.

Stakeholders

- Are the stakeholders at multiple levels—federal, state, and local?
- Is involvement from the public, the media, and lobby groups required?

It is important to involve all stakeholders in the process so they "buy in" to your solution, but the greater the number of stakeholders, the higher the risk. These risks can be mitigated by the negotiations and agreements during the planning process.

Overall complexity

- What is the overall complexity of the project?
- Are there federal regulations you must meet?
- Are multiple vendors involved?

Complexity in your implementation increases the amount of time you need to thoroughly deal with each issue.

Project planning

- Is your project planning well-defined?
- Is your implementation strategy consistent with the existing business strategy?

If the objectives of the project are not well-defined, you may spend large amounts of time and money on the wrong things. Following the recommendations of this course will help minimize these risks.

Project management

- Are you using proven methods?
- Is there built-in accountability?
- Is there built-in quality control?

Project scheduling

- Are scheduling deadlines reasonable?
- Do you have project-management tools to identify project milestones?

Project management and project scheduling are necessary to keep your project on time and within budget.

Project resources

- Do you have adequately trained staff?
- Is there a knowledge gap in your organization?

If you do not have adequately trained staff, you will need to develop a plan to acquire or train them.

Step 2: Discuss the nature of the risk in the context of the planned implementation

After identifying the risks, you must discuss the nature of the risk within the context of the planned implementation. A common risk associated with the implementation of new technologies, for example, is the existence of a knowledge gap. The new technology should, therefore, be discussed in relation to your staff's existing skill level. This type of discourse will lead to a better understanding of the level of risk faced, and how the risk can be mitigated.

Step 3: Describe the mitigating factors that will be used to minimize the risk

Once the risk has been identified and discussed, describe the methods to mitigate the risk. For example, the following measures could be taken to mitigate the risk of creating a knowledge gap:

- Generate an assessment of the current staff's skill level. Use the findings to develop a training program.
- The software will only be purchased from companies that have an established training program for their software.

In this case, two provisions were made to reduce the risk of implementing new technologies with an untrained staff.

Step 4: Assess the likelihood and seriousness of each risk and give it a score

Quantify each risk to your project based on its likelihood and seriousness. To quantify the risk associated caused by a knowledge gap, for example, you could ask the following questions:

- How likely is it that your staff will be faced with a lack of skills when confronted with the new system?
- What impacts will lack of individual skills have on project implementation and how will its occurrence affect the agency/department?

If you have addressed your staff's skill level, you can address the likelihood of the risk. You can determine the seriousness by relating the risk to the impact it will have on your system implementation and organizational functionality. Assign each risk factor a value such as high, medium, low, or a numerical weight.

Chapter eleven

Summarize the level of risk

The final step is to summarize the risk factors. All of the scores produced in step 4 are summarized and an average score is derived. The results are then used to determine if the project's risk level is acceptable to the organization. Major organizations may have their own approaches to risk analysis. In all cases, follow the applicable organizational methodology, as it is the organization that is taking the risk.

Make an implementation plan

"*The implementation plan should illuminate the road down the path to GIS success.*"

Make an implementation plan

You are now ready to develop an implementation strategy. This is a critical juncture in your efforts because implementation is when the effects of the underlying change (which can be profound) come into effect. This is when the messiness of the real world intrudes on your so-far orderly planning process. Only through a clear-cut path towards final implementation can we expect our efforts to survive.

Organizational issues

As in all stages of the planning process, management and organizational issues come into play at the time of GIS implementation. In most cases, your GIS will require data from multiple sources, which leads to the need for a systematic data sharing relationship with one or more other departments or agencies. Once data is developed, it is important that each organization or department properly use it, which adds the burden of creating clear user guides and metadata to the planner's to-do list.

Working with multiple agencies increases the complexity of the implementation, but it may be worth it to create a richer database and a more utilitarian system that will get used and allow further GIS applications. At this stage in the planning process, some contact with other agencies and partners already may have been made. Armed with your clear understanding of the information products that are to be created and the data requirements of your GIS, now contact each and every one of the organizations or departments that have to be involved. For example, in the case of state or municipal organizations, these contacts might include local government organizations, state organizations, federal organizations, partnerships, and multi-agency bodies

These relationships will all be different. Some will be more formal and some may even include legal dimensions. For example, a city planning department might be legally required to report all new streets immediately upon their opening to the keepers of an emergency route mapping application. Informal relationships might evolve with other organizations sharing common causes or purposes, such as a network of conservation groups agreeing to share environmental data to their mutual benefit.

You must consider how the relationships that your organization has with these disparate external partners will impact your GIS project. Ask yourself the following questions:

- Who among these partners might hinder or prevent your system implementation (i.e., are there any showstoppers)?
- Who needs to be kept informed about your project to ensure that you have continued support?
- What will happen if another organization fails to maintain its interest or commitments to your GIS project?
- Who is responsible for coordinating and managing the relationship?
- What would happen if the relationship ended?

To succeed in your GIS efforts you must involve all the stakeholders in the planning and implementation process. A stakeholder is any person, group, or organization with a vested interest in your project now or in the future. These stakeholders could include organizations with which you have interactions, as well as the end users of your planned GIS-related services (the public citizen or the business customer), the media, and lobby or special-interest groups.

Often during the planning process, organizations discover that other organizations have data previously unknown to them or information that they consider of value. For example, in Australia, six or seven state government agencies were found to have data that was useful to one another. Before GIS planning began, there had been little or no sharing of data. During the GIS planning process, they recognized that by simply sharing the data they already had developed with one another, they could produce new and better information products for reduced cost. In other words, sometimes creating interagency relationships can yield substantial financial benefits for all parties.

Since data exchange can reduce the costs of data acquisition (typically one of the biggest costs of a GIS implementation) there is a strong spirit of data sharing particularly among U.S. federal and state agencies. If your plan calls for data that is fundamental to your GIS to come from outside your own organization, you should enter into some kind of a formal agreement (such as a memorandum of cooperation or a legal contract) with the other agency. This agreement should address:

- What will happen if the other organization fails to supply its portion of the data?
- Who is responsible for correcting and updating data in a timely manner.?
- What backup and security is in place to ensure continued access to the data?

Also consider who will decide what further data is collected and shared, who will fund the data gathering (including maintenance and updates), and who will be responsible for coordinating the regular data administrative tasks.

There are also important technical issues related to data exchange that can trip you up if ignored: including the data format and accuracy, the metadata standards, and the data's physical location.

Two key questions should be answered to identify how multi-agency data exchange impacts your project:

1. Which information products require multi-agency data?
2. Is the benefit that will accrue to the agencies from the new information products the reason for multi-agency planning?

If the answer to the second question is yes, you have a strong reason to cooperate on data exchange, and you might also have the leverage to request joint funding of the planning process.

Legal issues

People are not perfect and neither is the geographic data they produce. In reality inaccurate or poorly documented spatial data does exist; the trick is being able to realize it and avoid using it. Such data may result from faulty data input methods, human error, or incorrectly-used programs or measuring devices. Some errors are more serious than others, and the resulting liability can impact other applications. GIS managers need to know the extent of their liability.

For example, in the field of surveying, original paper maps may have to be retained as a legal requirement. To date, digital data does not have the same legal status as the written record, and in the event of a legal dispute, it is the paper map that will be used. This may change as surveyors adopt new technologies.

During preparation of your implementation strategy you should check with the legal experts closest to your situation. The rapid acceleration in the use of GIS data by government and other institutions is raising many new and interesting legal questions related to the transition of maps from paper to digital forms. A complete discussion of these issues is not practical within this book, but it is important to realize that these legal issues are out there.

System integration issues

The GIS that you propose to implement in your organization almost undoubtedly represents new technology and processes in your organization. Very rarely will your planning process lead to a system that is completely new from the ground up. Instead, the typical GIS plan must consider how the GIS will integrate with the existing computing environment, the so-called legacy systems. As their name implies, legacy systems carry with them a history of usage and people will continue to rely on these systems. Not playing well with the legacy systems is seen as bad form. As tricky as it can be, clean connections to legacy systems are crucial to overall project success.

Written organizational standards and policies on system-integration issues related to the GIS will help smooth out the implementation process, reduce resistance to a new computing system, and build support among end users. Careful justification of any needed upgrades or changes to hardware and software already in place is important. Consider the following aspects of the existing computing environment in which your GIS must co-exist:

Current equipment: list the current vendor platforms in use across your organization (including remote sites).

Layout of facilities: gather or create diagrams of facility locations and their associated network facilities (by department and site).

Communications networks: list out all the types and suppliers of networks intended for use by the GIS link (dedicated, in-house, commercial). Understand what's happening on the LAN (local-area network) and the WAN (wide-area network). What are the protocols: TCP/IP, IPX, NFS? What are the bandwidths for each link that will be utilized by the GIS (e.g., T3, T1, ISDN)?

Potential performance bottlenecks: identify missing or inadequate communication links, fault tolerance, system security, and response-demand issues.

Organizational policies and preferences: the information technology (IT) culture of your organization should be considered. Is all hardware from a single preferred supplier? Is there a standard operating system in use? If policies for procurement and standards for use have been adopted or maintenance relationships established, consider these carefully. Adopting different systems may lead to problems with support and acceptance and require additional staff training.

Future growth plans and budget: the GIS must work within the fiscal framework of the organization. Budgets set must be adhered to, so the GIS manager must always keep an eye on costs and making sure the work is efficient and affordable. If money runs short, things like breadth of user applications, system performance, or reliability will take a hit. Initial budget projections should be considered in later benefit-cost analyses.

Keep in mind that with the introduction of any new information system into an existing operation, there are bound to be bumps in the road. Even if you adopt industry-standard hardware platforms and use only well-tested GIS software, your application programs may still have bugs during their first releases. Long term GIS managers consider that they manage three interconnected items: budget, schedule, and functionality. Any two will have an impact on the third. You should prepare users for this to manage expectations while at the same time letting them know that their clearly communicated feedback will help you fine tune the system into well-oiled machinery.

Request for proposals

An RFP is a request for proposal that invites vendors to propose the most cost-effective GIS solution to a specified business need. Well-crafted RFPs are not shopping lists of hardware and software to be procured, but rather descriptions of the work that has to be done by the system. An RFP also specifies how the selection and actual procurement will be carried out so that the vendors who wish to compete can propose realistic options based on the most complete information.

Sometimes the actual RFP may be preceded by something called a request for information (RFI), which many organizations use to solicit assistance in the development of the final RFP. Savvy technology managers in organizations realize that early input from tech suppliers can be very instructional, because it is the technology vendor who spends the most time thinking about how to implement their particular technology. Sometimes an RFI is nothing more than an early draft RFP. RFIs are particularly useful when the scope of your system is likely to be large. You can send the RFI to vendors and ask them to respond with a letter of intent to bid and any comments on the included draft RFP. This step helps to design an RFP that is not inadvertently biased to one vendor or another.

The RFP for a major procurement is typically a single document supported by substantial appendixes. For a smaller project, the single document may be all that is needed. The document will lay out the requirements and the appendixes will provide details such as the master input data list, the information product descriptions, copies of government contract regulations, worksheets for product cost and data conversion estimates (to standardize replies from vendors), and overall data processing plans pertinent to the GIS. The main document of the RFP should include the following:

General information/procedural instructions: this is where you cover the acquisition process and schedule, the vendor intent to respond, instructions on the handling of proprietary information, proposed visits to vendor sites and debriefing conferences (if deemed necessary), clear rules for the receipt of proposals (date, time, place), and, finally, arrangements for communications between the contact person in the organization and vendors.

Work requirements: this section is the most substantial part of the RFP. It should list the work to be done by the system, expressed specifically in the context of the information products that have to be created, the system functions that are required to produce the information products, and the data that needs to be in the database. The main text of the RFP should list what is required with supporting definitions and descriptions as necessary in the appendixes.

Here's a checklist of work requirements that every good RFP should include:

- The full set of information product descriptions
- The master input data list
- Data-handling load estimates over time by location

- Functional utilization estimates over time by location
- Notes on existing computer facilities and network capabilities
- Lists of any special symbols that are required

Services to be performed by the vendor: vendors are typically contracted to supply and install the hardware and software systems and train the new users. Clear, complete written documentation in the form of user guides must be created. After installation and start-up training, also plan for ongoing maintenance and upgrades to the system.

References: good vendors are more than happy to provide reference sites of existing customers running similar systems. Don't only ask for the references: call them yourself and hear their experiences.

Financial requirements: some of the significant financial decisions that will be required involve whether the actual hardware equipment will be leased, lease-purchased, installment-purchased, or purchased outright. Depreciation and tax considerations may come into play. Maintenance on software is a guarantee of prompt upgrades to new versions and technical support from the vendor.

If estimates of operating costs are required from the vendors, staff hourly rates will need to be shown. The RFP should include the instructions on use of financial worksheets, if any, the licensing terms and conditions on purchased data and software, the cost of the recommended approach, and several alternatives and their respective costs.

Proposal submission guidelines: you need to explain the required format, the number of copies needed and list of contents of the proposal including appendixes. Specify any mandatory contractual terms and conditions, or any specific financial requirements.

Proposal evaluation plan: this is the part of the RFP that details the structure of whatever committees you will use to evaluate the proposal. It steps through the evaluation process and lays out the actual evaluation criteria that will be applied in making the final decision. These are the rules of engagement, as it were. Care should be expended to spell out as clearly as possible the process so that all minds are in agreement within the organization and the vendors on the process that will be followed. If the system will be subjected to benchmark and acceptance testing, the testing processes should be provided. This is particularly important in major procurements; most organizations contemplating major procurements have firm procedures for these steps. It is in the interests of both the organization and the vendors to produce a clear, explicit RFP.

Security review

System backups and data security are designed to protect your GIS investment. One of the major benefits that come to organizations is the increasing value of their databases as they grow over time. Whatever the current value of your database, if it is properly maintained, its value in five years will increase dramatically. The successful GIS will often quickly become an integral part of an organization's daily operations. So over time the value of information derived from the GIS database increases due to the improved business processes the GIS functionality delivers. These are the reasons why a sound security plan is required.

The risk of your system or data being destroyed or somehow compromised is real and deserves serious attention. Any information system is vulnerable to both deliberate and accidental damages. A disgruntled employee might purposely corrupt data, hackers may steal information, or a computer virus could find its way into the server through e-mail. Natural disasters also pose a threat. Earthquakes, floods, fires, hurricanes, tornadoes, and lightning are all examples of natural hazards that could disrupt a GIS.

While deviant behavior and natural disasters are intriguing subjects, threats more common to many organizations are found in day-to-day operations. Consider the potential effects as coffee spilt in the wrong place, a well-intentioned employee who accidentally deletes or corrupts a database, or a power disruption with no automatic battery backup.

When conducting a security review, examine the physical, logical, and archival security of your databases.

Physical security measures protect and control access to the computer equipment containing the databases. Physical security guards against human intrusion and theft (security doors, locking cables) and environmental factors such as fire, flood, or earthquake (fire alarms, water proofing, power generators).

Here are some recommendations for physical security:

- Restricting access to the room in which main data storage terminals are located
- Reviewing construction plans for new stations as available
- Installing fire and intruder protection alarms
- Implementing document sign-out and follow-up procedures

Logical security measures protect and control access to the data itself either through password protection or network access restrictions. A common security measure is to specify that only database management staff have editing and update rights to particular data sets. Here are some additional ideas on logical security:

- Develop a policy for terminal access
- Protect and control all storage media
- Develop a schedule for virus scanning
- Create a matrix showing access by document type

Archival security means ensuring that systems are backed up and that these backups are stored correctly in remote off-site locations. Legally many organizations are required to archive their data. This means that the system must include functionality that creates these archives. Raw data transfers are not enough: metadata, information about past coding and updating practices, and the location of data must all be stored to allow for quick recovery in the event of a system failure. Consider:

- establishing an off-site facility to store archived data
- establishing an audit trail to track copies of data sets
- capturing every transaction

After your initial security review, address the current and future security risks and make recommendations. These recommendations should cover physical, logical, and archival issues related to security. The table below lays out what a typical security review might look like:

Physical security	Logical security	Archival security
Prevent access to main data storage from the back stairs.	Develop a policy for terminal access.	Establish an audit trail for copies of data.
Review the construction plans for the new building to ensure appropriate climate control.	Create an access matrix by document types.	Establish an off-site backup facility.
Build a public workroom so staff don't have to go into the vault room to do their work.	Review protection of storage media.	Create and organize metadata.
Initiate document sign-out and follow-up procedures.	Purchase antivirus software.	Purchase storage media.

Chapter twelve

Staffing and training

It is hard to overemphasize the degree to which a successful GIS is dependent on the staff who will build it, manage its evolution, and maintain it over time. The best GIS plan in the world will not launch a successful GIS: that takes people.

There are some basic staff positions required for a GIS implementation that all have associated skill requirements. In particular, this section discusses these staff requirements, looks at the GIS "knowledge gap," and the training options for getting GIS staff up to speed.

The number of staff associated with a GIS depends on the nature and size of an organization. In larger organizations with widespread user groups and more complicated applications, a wide range of staff could be required—everyone from network and database administrators to GIS specialists and programmers and of course managers to oversee the work of all these individuals. In smaller organizations, the staff may have to wear many or all of the hats, but that can also work. For example, as the sole GIS manager you may have to take charge of all the GIS planning and the system design and administration. A lone GIS analyst may have responsibilities spanning different software packages and applications.

As previously noted, the first line of division breaks between the core GIS staff and the GIS end users. Crucial staff, but not really part of either of these two groups, is the management group and system administration team. These two groups require special attention as you set up a working team. Let's consider them further.

The fundamental purpose of a GIS is to provide to management new or improved information for decision-making purposes. At this level, the managers actually become customers whereas up until now they have been involved in helping you pass through the hurdles of the organizational bureaucracy. The managers could be the project sponsor, a member of executive management who will be using the application (or its output), or a management representative who can serve as the conduit delivering management's requirements.

GIS staff

The GIS staff includes everyone directly concerned with the design, operation, and administration of the GIS. The GIS staff includes:

GIS manager: the GIS manager requires skills in GIS planning, system design, and system administration. Ideally the manager would also possess some of the hands-on technical GIS skills. In fact, most effective GIS managers do come from the ranks of the GIS doers. The responsibilities of this position vary depending on the type of organization. In smaller organizations the GIS manager may be the only person involved with the GIS so the same person who negotiates data sharing agreements with the next county over is also pushing the buttons on the GIS when the data itself shows up. In very large organizations the GIS manager is often in charge of coordinating the GIS staff, working with multiple departments, and overseeing the development of an enterprise-wide database. An emerging trend is towards the creation of geographic information officers (GIOs), who become the true champions and executors of GIS innovation within the organization.

GIS analysts: GIS analysts are persons with GIS expertise working in support of the GIS manager. In smaller GIS organizations the GIS analyst may be one person with a broad range of GIS skills. In large organizations there may be several GIS analysts. Within the ranks of GIS analysts you might find titles, including:

- The GIS technology expert is responsible for the hardware and network operations of the GIS.
- The GIS software expert is responsible for application programming.
- The GIS database analyst is responsible for administration of the GIS database.
- The GIS primary users.
- The professional GIS users support GIS project studies, data maintenance, and commercial map production.
- The desktop GIS specialists support general query and analysis studies.

211

Chapter twelve

GIS end users

It is helpful to think of the end users of your GIS as an important staff component because they affect the design and utilization of the GIS. GIS end users may include:

Business experts—these are the key employees with intimate knowledge of the processes that your GIS is attempting to improve. Expect them to provide major input to the GIS manager regarding the design and management of the GIS.

Customers—these are the clients of the GIS who are served by it and would include business users who will require customized GIS information products to support their specific business needs. They range down to the more casual Internet and intranet map server users, people who will access basic map products invoked by wizards or Web browsers.

System administration staff

In addition to the GIS staff, a large organization will have a system administration staff. This staff plays an important role in the day-to-day operation of the computer systems that will support the GIS. Members of the system administration staff may include the network administrator responsible for maintaining the enterprise network, the enterprise database administrator responsible for the administration of all databases that interact within the organization, and the hardware technicians who handle the day-to-day operation of computer hardware within the organization.

Clear job descriptions are essential before hiring begins. Outlining GIS roles provides you and your prospective employees with a common understanding about the position and its requirements. Job titles and descriptions will also come into play during job evaluation and performance reviews.

If your organization is new to GIS, the human resources department will not have suitable job descriptions for the type of GIS personnel you need. In this case you'll need to write the job descriptions yourself. Be sure to design them realistically. Obviously, you want the best person for the job, but if you require that your digitizing technicians have master's degrees, you will probably never find a digitizing technician. Refer to the sample GIS job descriptions in appendix A.

Once you've identified the GIS staff that is required, you must decide where the positions fit into your organizational structure. There are three main levels in which GIS staffs are typically placed. The decision impacts the role and visibility of the GIS department:

1. Within an existing operational department: in this scenario, staff is tied to a specific need and budget. It is difficult for staff at this level to serve multiple departments.

2. In a GIS services group: this group serves multiple projects, but still has the autonomy and visibility of a stand-alone group.

3. At an executive level: this placement signals a high commitment from management. Staff have high visibility and the authority to help coordinate the GIS project. The downside is when GIS staff at the executive level become isolated and foster the perception among the rank-and-file that the staff is perhaps out of touch with some of their critical stakeholders.

4. In a separate support department: the more old-fashioned notion of IT would place any new information system staff in a centralized computing services department. This is the "systems" group in many organizations.

Remember that a good staff is an invaluable tool to a manager. Money can buy more hardware and software, but even money cannot create the motivation and enthusiasm that is essential to a successful staff and a successful GIS implementation.

The knowledge gap

The term "knowledge gap" describes a phenomenon that results from technology capabilities growing faster than an organization's ability to utilize the new capabilities. In the case of GIS technologies, there was no knowledge gap before 1990: people were still more capable than their GIS systems.

By 2000, the rate of GIS development was above the normal growth trend of institutional management skills. This means that systems are now more capable than people and the normal growth in skills within an organization does not keep up with developments in technology. The development curve of GIS has leveled off somewhat recently, but management still has a lot of institutional learning to do before the full capabilities of GIS are truly utilized. The knowledge gap has clear implications within an organization's training strategy and budget commitments, but it also bears on the ability of higher learning institutions to provide some of the needed skills training. The degree to which an organization can improve its collective skills base will have a direct bearing on how well and how far the technology will be adopted within the organization. These skills are the tools that an organization uses to actually leverage technology in order to increase productivity. Only by recognizing that that there is a knowledge gap and implementing efforts to fill it in can one realize that the knowledge gap is also an opportunity gap.

Some organizations attempt to circumvent the knowledge gap by hiring consultants to operate the GIS under the assumption that these hired guns will bring the needed knowledge into the company. While this is certainly a possibility, there are few instances where it can be fully recommended. If you can afford to hire a consultant, you can probably afford to hire staff. Hiring a consultant is just another way of buying the skills that you really need in your organization.

Consider the tale of a Canadian oil company that spent $200,000 implementing a complex environmental GIS application. When the application was finished, the two key staff members left, and there was no one left at the company who could use the system. Everyone involved in the planning and implementation must have felt like they'd been wasting their time. The moral of this story is that an organization needs smart people in place who can really use the system. Building a strong internal staff is a process not an event in most cases. Highly selective hiring practices and ongoing training are a must. You must also create conditions that are favorable to learning and foster independent thought. Provide interesting and challenging tasks, a supportive management environment, and continuing opportunities for knowledge development (including training and formal interaction with other GIS professionals).

Training

One of the primary causes of underutilization of GIS is lack of staff training. It should go without saying that if people cannot make the system do what they need, they'll quickly abandon it and stick with the old safe way of doing things. And who could blame them. They have a job to do, and the mere existence of a fancy new GIS system is not going to help them or run itself. It takes a live, thinking human being to frame a spatial problem in the context of a GIS. In order to get meaningful answers to their questions, they need to know how to apply the tools to the work as they understand it. Train, train, and then train some more.

Consider how the people within your organization will use the GIS before developing a training program. GIS staff and GIS end users will require different types of training.

Core GIS staff training

The core GIS staff is the cornerstone of your efforts. They will require up-front and ongoing training to keep them current on new techniques and methods. The staff is responsible for creating, maintaining, and operating both the data and the system infrastructure. In addition to the GIS manager, a full GIS staff for an enterprise-wide GIS includes GIS analysts (GIS technology experts, software experts, database analysts, and primary users) and GIS technicians (hardware technicians and cartographic technicians). In addition to the GIS staff, you may also need to consider training for system administration staff (network administrator, enterprise database administrator, and hardware technicians).

The training program developed for the GIS staff could involve courses in database management, application programming, hardware repair, or even geostatistical analysis depending on their relative and collective skill sets. The training your staff receives should complement their job responsibilities. These are the people responsible for maintaining a product for your users. A well-trained staff is crucial for the continual success of a GIS.

GIS end user training

The training required for GIS end users is understandably much less involved than what is required for the core GIS staff. Often a single interface or Web application is all that an end user will ever see, meaning the application can be taught in as little as minutes.

Many vendors provide training courses related to their own software, and some even cover basic GIS theory and applications as necessary.

Chapter twelve

Self-study workbooks that include software to practice with provide another flexible learning alternative. GIS is inherently a multidisciplinary endeavor, so training in other areas beyond the actual software continues to play a major role in developing thinking GIS managers.

GIS manager training

Ideally the person hired as GIS manager will have demonstrated GIS technical competency. But these skills need to be continually updated if the manager is to be able to give meaningful direction to his or her staff. It is also important that GIS managers demonstrate or work to acquire effective management skills. Courses designed to help in specific areas such as general management skills, project management, strategic management, and total quality management will all be helpful to a GIS manager (or any manager for that matter).

Training delivery

Training can be delivered in many flexible forms these days thanks to wonders like the Web and cheap air travel. Face-to-face classroom courses are available from vendors or educational establishments at their premises or on site with you. Web-based training, distance learning, and self-study workbooks are all options for training in GIS and related areas. Whatever method is used, sufficient time and resources for training and related activities (travel, preparation of assessment, follow-up reading) must be provided.

Organizing your project

Before starting a GIS implementation, it is important to consider how you will organize your relationship both to management and GIS staff. Clarification of roles and responsibilities and establishing good lines of communication will greatly aid the implementation of any GIS. The GIS steering committee and the system development team are the key players in this function.

A GIS steering committee helps you make decisions regarding project management, changes in project scope, and what to do if your project gets behind schedule. It is the fundamental link to senior management in your organization. The GIS steering committee should take an active role in the GIS decision-making process. The table below shows responsibilities of the GIS steering committee.

GIS steering committee responsibility	Description
Review the system development team reports.	GIS steering committees should meet monthly to review reports from the system development team. It is important that the steering committee be actively involved in the decision-making process.
Review GIS project status.	During the steering committee meetings, the current status of the GIS project relative to the implementation milestones should be reviewed.
Identify and give early warning of problems.	Reviewing the progress of the project compared to its schedule will help to identify problems as early as possible and allow adjustments to be made.
Make necessary adjustments.	Every time a GIS project requirement changes or something needs to be added or deleted, the change should be documented. If you can manage a project within the scope that was already identified, the likelihood of finishing it on time is good. One way to manage the scope is to put the steering committee in the position of determining when you will expand a project and when you will not.

The steering committee can also decide when you need to reprioritize or scale down a new application or project extension. The makeup of a steering committee will vary according to an organization's nature and size. At large sites, the steering committee could include the project initiator (known as the project sponsor), the GIS manager, at least two customer representatives, a management representative, and a technical expert to reinforce the technical perspective of the GIS manager.

At medium or small sites, the steering committee could include a project sponsor, the GIS manager, one customer representative, and management representative.

The system development team oversees design implementation, reviews progress, and identifies problems as they arise. The team is responsible to the project steering committee. The system development team can be as small as one or two people at a small site or an initial four or five people

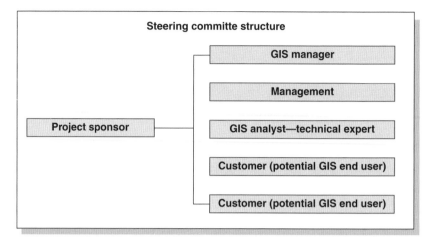

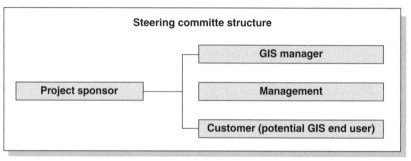

at a medium site (it will tend to grow). Whatever its size, the tasks that have to be accomplished by the system development team are consistent. It is important to include the customer in the development process and perform. The team's primary functions are shown in the table on the following page

As mentioned above, the makeup of the system development team will differ depending on the size and nature of the organization. In all cases, the system development team should include a GIS manager. At large sites, the system development team could include the GIS manager, an expert from the business side of the organization who thoroughly understands how the application helps, a GIS technology expert, and a GIS database analyst who understands the logical data model and linkages between the different data elements.

At small or medium sites, one person may do many of the jobs required, but it is important that this type of expertise come together as a system is being developed. At such sites, the system development team might

System development team responsibility	Description
Solve design problems.	The system development team discusses and resolves design problems as they arise.
Focus activities on the critical path.	The critical-path method is one of a number of project-management tools that can be used. It identifies the tasks and the sequence of tasks most critical to complete the project on schedule. Noncritical tasks are also identified. These are tasks that can be delayed without affecting the overall project schedule. The critical-path method is often useful for focusing the system development team on the most important activities in the development of the project.
Balance work-load assignments.	The system development team is responsible for balancing the workload and assignments given to staff.
Meet regularly and report progress.	The system development team prepares periodic progress reports (usually monthly) for the steering committee. The system development team should have regular development meetings. Initially, these might be daily meetings, then be held weekly. You should set up a formal schedule and reporting process where every week the team meets to at least touch base. Often, a programmer will make some assumptions about requirements and then spend one or two weeks creating something that does not work. Frequent meetings are a way of trapping problems such as this and getting back on track quickly.

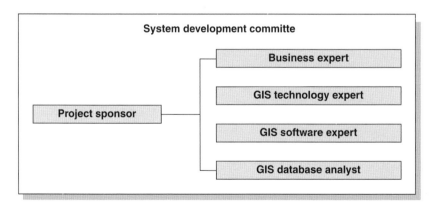

include just a GIS manager, an expert from the business side, and a database analyst.

The acquisition plan is in one sense the final step in the planning process, because after it's approved, you move into actual implementation. Now obviously the planning for GIS continues as long as there is a system to operate, but that original window of planning before a project launches is now shut. From here on everything happens in real-time in a production environment. The acquisition plan will become your guide to implementation. Once it has been thoroughly reviewed, it is presented to executive management for their approval and funding of the project. After you receive approval, you are ready to start acquiring and implementing your GIS.

The acquisition plan should do several things very effectively:

It should recommend an actual strategy that outlines the specific actions that will be required to implement your GIS plan. It should review all of the previous planning work and include all of the relevant documents to justify your recommendations. It should highlight the latest (if any) extra implementation actions or special concerns that have been identified. Finally, it must include a list of actions and concerns that are presented to senior management in the final report. Here's a review of the key components of the acquisition plan:

GIS implementation strategy recommendations

To develop strategy recommendations for your GIS implementation, you review each step of the implementation strategy discussed in this chapter. You need to create recommendations for each component. When you have finished you will have a list of the actions that are required to implement each component of your plan.

After completing your review you should also note any extra implementation actions that are now required or any new concerns that you have identified during the review process. Add them to your final recommendations.

Institutional interaction requirements

If there is a need to interact or cooperate with multiple organizations, agencies, or departments in the course of your GIS project, you should recommend that agreements relating to the responsibilities of those involved are necessary. Such agreements should, at a minimum, be documented in a memorandum of understanding (MOU). If the system depends on these interactions, formal contracts are required. As enterprise-wide GIS systems become more common, it is important that the data requirements of all the organizations involved are included in the planning process.

Legal review

A legal review should be recommended when there are legal issues associated with the use of data in your GIS. For example, there may be a requirement to preserve the original paper documents containing legal descriptions for land parcels in the United States even though these may not be used again after automation.

You should also recommend measures to reduce the risk and liability associated with possible errors in your organization's GIS data and information products. There are various types of error in data and it is important to understand your organization's liability and recommend actions to limit that liability.

Existing computing environment

The GIS hardware needed by the new system has been outlined, in conceptual form, as part of the implementation strategy. Your organization may, for example, have significant time and money invested in the existing technology base. Your implementation strategy recommendations should take your organization's hardware and software preferences into account.

System requirements definition

The system requirements definition should be clearly spelled out within the requirements section of the request for proposals described in this chapter. It lists the work to be done by the system, including the information products that have to be created, the system functions that are required and the data needed in the database. You must recommend the actions required to put the system in place.

Security review

You have conducted a security review to determine the type and amount of security necessary to protect your GIS from damage. Through an appraisal of the security review, you can make recommendations on how to reduce the chance of system damage and data corruption.

Department staffing

Staffing issues have been discussed with each department affected by GIS implementation. Staffing a GIS is a long-term operational cost and a major expense for most enterprise-wide systems. You should include recommendations about the number and types of staff required to implement and maintain the system. This also shows the committee that you've done your due diligence and communicated as widely and clearly as you could.

Training program by personnel category

After determining the staffing needs for your GIS implementation, assess the GIS capabilities of your existing staff and determined an appropriate training regime. Keep in mind that continuous employee training will be necessary for all GIS staff and must be included in your budget. Recommendations for an employee-training program that provides for the necessary levels of GIS staff should be included in the implementation plan.

Cost model

You have developed an overall cost summary for your system. This was done to allow you to analyze the benefit-to-cost ratio of your organization's system (see chapter 11). All the costs expected for GIS implementation for the planning period are documented within a cost model. The cost model accounts for five cost categories: hardware and software, data, staffing and training, application programming, and interfaces and communications costs.

Benefit-cost analysis

Benefit-cost analysis should also be included in the acquisition plan (see chapter 11).

Migration strategy

You have examined the existing computing environment of your organization. Now you must recommend a migration strategy for moving from the existing system to the new GIS. Include with this strategy detailed plans for merging the new system with legacy systems. Recommendations should address whether or not to replace, rebuild, or merge the new system with modeling techniques used by any legacy systems.

Alternative implementation strategies

Alternative implementation strategies should be included in the implementation plan. Recommending more than one approach (even if you have a strong personal favorite) shows that you're looking for the best approach for your organization. Consider using pilot projects with a subset of the geographic extent, particularly if you're concerned about the ease of integration of any particular data sets. Beware, however, of using a pilot project as a replacement for planning. This approach will inevitably lead to an incomplete system design and a certain measure of frustration and wasted time.

Risk analysis

Implementation can be impacted by four major groups of risks: technology; budget constraints; project management and scheduling; and human resources. Recommend steps to mitigate the risks identified. Include the results of the risk analysis in the final report to help ensure that senior management is aware of any implementation difficulties that were identified.

GIS steering committee review and approval

At this point in the planning process, your implementation strategy should be reviewed and approved by the steering committee. Never rely on a consultant to undertake this review. It must be undertaken by someone inside, and preferably by a group and not an individual, to reap the strength of combined views.

The GIS steering committee should already be fully aware of the information that has been prepared during the planning process. This information is the basis for your implementation strategy. They should now thoroughly review all the planning materials, and then approve (or revise, or reject) the recommended implementation strategy. If you've done your homework, the approval should come quickly and with renewed enthusiasm for the mission.

The GIS steering committee can, of course, adjust the implementation plan. They should also provide ongoing support during and after implementation. This committee is an integral and permanent part of the GIS team effort. The members might change, but the committee will always be needed.

System procurement

As part of the implementation plan you must consider how you will procure your system. Two key factors will affect how you go about it: the procedural requirements of your organization and the characteristics of the planned system.

Many organizations have purchasing requirements that become more rigorous as the expected expenditure increases. Low-cost, generic products can often be purchased without a formal process. Acquisition of more expensive products and services may require most, if not all, of the following steps. The following list of steps represents about the most elaborate procurement process you could encounter. The average organization might include only half of these steps:

Step 1: Request for qualifications (RFQ)

The RFQ is a request for qualifications from each technology vendor. The RFQ is optional and done early in the procurement process. The response should identify the types of systems that a vendor has provided along with a track record of experience.. This will help determine which vendors have experience in providing systems similar to the one that you plan.

Step 2: Request for information (RFI)

The RFI is particularly important if you have a large procurement that could be so coveted by vendors as to result in potential protests from vendors.

Step 3: Request for proposals (RFP)

The RFP asks vendors to propose the most cost-effective combination of hardware, software, and services to meet the requirements of your organization.

Step 4: Receipt and evaluation of proposals

Identify who in your organization will serve as evaluators and the evaluation criteria that they'll be using to make that evaluation. These criteria are often a key part of the RFP.

Step 5: Benchmark test

Benchmark testing involves verification of the proposals from the top-rated vendors to ensure they can perform the tasks and functions necessary for the planned system and communication network. A section on benchmark testing is included as an appendix to this book. Vendors may not wish to participate in a benchmark effort if you request extensive testing, but are not planning to make a relatively large acquisition. In the event that thorough benchmark tests are not carried out, the purchaser should request a demonstration of system capabilities that includes the full functionality required for the organization. Otherwise its caveat emptor, let the buyer beware.

Step 6: Negotiation and contract

Negotiation and contract often involves cooperation with your purchasing and legal staff. Personnel from these two departments can really help during contract negotiations, but just make sure as the manager you review the contract to ensure that the technical requirements are properly addressed.

Step 7: Physical site preparation

Make sure that the site is properly prepared before the system is installed. Seems like an obvious point, but it can mess you up if overlooked. Are the necessary servers and appropriate network connections available for system installation? Is the computer room properly ventilated? Does each GIS team member have a chair and a desk?

Step 8: Hardware and software installation

Specifying who will install the system's hardware and software depends on the complexity of the system. In some cases, the vendor will provide this service. If your organization has the technical staff available, you could install the system in-house. Basic PCs have evolved a long way. It's possible to take delivery of a boxed computer and have it set up and running in less than hour. Often, the vendor and the client are both involved in system installation.

Step 9: Acceptance testing

Within the RFP you would have specified the methods chosen for acceptance testing after system installation. These tests often require the hardware components to operate, without error, for a given period of time. With many of the widely used components available today (e.g., personal computers and their operating systems) these tests are of limited value. When planning acceptance tests, consider testing the system's ability to integrate with existing databases and software and test all these connections. Network connectivity problems are always a major source of bugs and should be tested hard for possible problems.

Selection criteria

The final choice of your GIS is based on a combination of its ability to perform the functions specified to create the information products. So this is the first litmus test of whether the system is acceptable. Some of the other factors that play into the selection criteria might also include: cost, training availability, system capacity and scalability, system speed, system support, and last but not least vendor reliability (i.e., their financial stability, position in the marketplace, and checked-out references).

Return to the preliminary design document and retrieve the total function utilization table. This table was created during the conceptual system design. The functions are already prioritized as well as classified based on their frequency of use and relative importance to total system functionality. The classes derived from this process are used in the selection criteria.

Finally, ask the vendors to put their proposals into a format that will allow you to develop a rating system for proposal comparisons. If you are procuring a major enterprise-wide system and benchmark testing is planned, let the vendors know that their proposals will face these tests.

Managing implementation change

The business model of the modern organization is characterized as dynamic. Implementing a GIS to fit this model is a continuing process of change since it occurs in an ever-changing environment. There are changes in technology, both in hardware and software. But there are also sometimes subtle, sometimes profound, changes in the business needs of your organization and perhaps in the institution itself. Managing change starts with understanding the types of change.

Technology change: if you don't think technology is changing GIS, consider that over just the past 15 years, the hardware used for GIS has evolved from mainframes to minicomputers, from minicomputers to workstations, and from workstations to personal computers. Operating systems have also changed from being proprietary and hardware dependent to hardware independent. To accommodate these changes, GIS software has constantly evolved.

The rate of technology change and the lifecycle of technology were introduced in chapter 7. Change is generally a good thing when it comes to technology. New versions of software and hardware do actually make the work more easier or more cost-effective. Newer hardware means significantly faster performance, storage is cheaper, and work is made easier. New versions of GIS software tend to offer easier use, simplified procedures, and reduced repetitive work. New versions tend to fix bugs, and are thus less error-prone and more stable and reliable. The rate of change in technical capability is swift. Advancements in technology move faster than the changes in institutional requirements and must be treated using the criteria of cost-effectiveness.

Rapid advancements in technology also mean that maintaining older software and hardware may become prohibitively expensive in less than five years. Vendors must offer maintenance for new hardware and software. Over time, they will increase the price for maintaining older technologies to focus more of their resources on the latest technology, encouraging the migration to newer versions.

Institutional change

The business needs of most organizations change incrementally over time. The staff gains more experience, new ideas and approaches are brought into the organization, and new information products to support the new business needs are requested. It is the natural growth cycle of a successful business operation. Incorporating the new information products into the GIS workload will present no significant problem if growth is planned for in advance.

Occasionally, major institutional changes occur, such as the merging of companies or departments of government, or major changes in mission for all or significant parts of the organization. In these instances it may be that one organization has done no GIS planning while the other is quite advanced. In any event, it is likely that a new overall suite of information products will be required involving new data sets and database designs. In these circumstances, it is recommended that the entire GIS plan be reviewed and new enterprise-wide planning be put in place, with new ramifications for benefit-cost, technology acquisition, and communications.

The rates of change in technical capability and change in institutional business needs are illustrated in the diagram on the following page. The differences in the rate of change are clearly illustrated. The changes in technical capability are managed on the basis of cost-effectiveness. The introduction of new information products in response to changing business needs is managed on the basis of maximizing support for the business

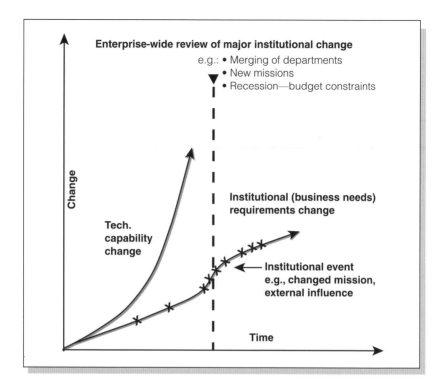

of the organization. The challenge for the GIS manager is to provide the maximum business support while minimizing expenditures.

In the early days of GIS, plans were typically updated on a five-year cycle. These days five years is too long; things change too fast. Determining when your implementation plan becomes outdated requires careful assessment of the technology changes that have occurred since the preparation of the last plan. When updating the implementation plan, the most important considerations are the need for the continued availability of existing information products in the most cost-effective manner, and the expansion of the information "product line" to meet the changing business needs of the organization.

Recommendations for managing change

Start with an enterprise-wide plan

The best advice for managing change is to start with the goals and objectives of the entire organization in clear focus. This planning approach ensures that the GIS integrates with the overall strategic objectives of the organization. Enterprise-wide planning allows for maximizing the benefit that can be gained from GIS. The most important system functionality will be identified on an enterprise-wide basis. The communication requirements of the GIS throughout the organization can be considered intelligently. The time spent in enterprise-wide GIS planning is the best investment related to GIS that you can make because it sets you up for success.

There are special circumstances where planning can and should be adopted for one or a small set of very high-priority information products. Things created in response to crises or natural disasters are an example. There are also examples of out-of-the-box software being used with available data to produce a useful information product (followed by another and another). But the demands and problems start to arise and you start to be forced to actually think about what you are doing. You need to plan within the normal life of an organization. The best advice is to plan early and as broadly as possible.

Note that enterprise-wide planning does not imply enterprise-wide implementation. In the end it may be prudent to implement only a small part of the plan at first and build on the success of that part for further implementation. Plan comprehensively and implement incrementally.

Adding information products

Adding new information products once the GIS is up and running is not a problem, but it does require careful management. Revisit your activity planning when adding a new information product. Determine the priority level of the new information products, their demand on the data-entry schedule and the impact on other information product generation due to displacement or delay. In essence, you must manage the process of scheduling information product design and generation in the context of the existing information product priorities, data availability and readiness, activity timing, and revised onset of benefit from other information products.

Information product design, particularly for more complex information products, can be a time-consuming process and must be judged in the context of the lifecycles of the technology being used. For example, if the technology currently used has one year until obsolescence, a major information product design should probably wait and take advantage of the new technology. This may significantly reduce the amount of time needed to develop the information product and reduce costs measurably. Alternatively, the original design could go forward but with the full understanding that its benefit cycle will be limited and that it may face abandonment or require a significant effort to adapt it to the new technologies when they are put in place.

Technology acquisition

The first recommendation concerning technology acquisition is to ensure that the equipment procured is sufficiently utilized from the outset. At least 50 percent of the equipment's capacity should be used in the first year of its acquisition. The old paradigm of buying the largest equipment possible from the current budget and planning to use 10 percent of its capacity in the first year, 20 percent the second year, 50 percent the third year and so on over a five-year cycle, is no longer cost-effective given the increases in equipment capacity and the cost reduction over each 12-month period.

It follows that it is cost-effective to have a continuous technology acquisition budget to maintain your system. Senior management needs to allocate resources to keep your system cost-effective. Hardware, software, training, and data all need to be procured bearing in mind the lifecycle both of the technology and the information products. To do this, describe your GIS as part of the organization's infrastructure. As with other elements of this infrastructure, two types of funding are necessary: 1. funding for on-going operation and maintenance, and 2. capital investment. A common approach to on-going capital investment is to estimate the lifecycle of the asset and annually budget for an appropriate amount in a capital replacement (or "sinking") fund. For example, three years may be the estimated lifecycle for a desktop workstation. If your organization has 60 of these workstations, you should budget funds to replace 20 per year. In common practice, hardware extends its useful life by being cascaded down through the organization, with older workstations being assigned to less-demanding tasks and the high-end workstations being replaced on the acquisition budget. Many organizations also regard data acquisitions as a capital investment item rather than operation and maintenance expenditures.

Management communication for change

Continual change raises a number of questions for the GIS manager, for example:

- How do you secure support for a technology that is constantly evolving?
- How do you convince management that system upgrades are necessary?
- How do you secure funding for training costs?

When facing these questions, keep in mind that GIS implementation will go ahead because your organization expects to receive a benefit from using GIS. Also, the technology and the business needs will change with time and both you and your senior management must be educated about these changes.

Keep senior management informed about the benefits received from the GIS as they become known. The support required for upgrades and training must be presented in the context of these benefits. The GIS must be shown to be minimizing cost while maximizing support for business solutions. Revisit the benefit categories in your benefit-cost model and identify demonstrable benefits that have occurred. Identify the policy changes that have taken place that have been substantially influenced by information received from the GIS. Provide these in a continuing series of reports to senior management when demonstrable benefit can be exhibited. Refer to these reports in your budget planning to maintain the cost-effectiveness of your system.

Keeping your plan current

Based on what you now know it should be clear that the GIS implementation plan requires continual review. It must be periodically updated in response to changes in your organization and to keep pace with technology trends. Determining when to update the GIS implementation plan requires a careful assessment of the changes that have occurred since preparation of the last plan. View the components of your GIS as assets that depreciate and that require a certain level of investment to stay cost effective. Minimize the total cost of the GIS implementation while maximizing support for business, and when updating the implementation plan the most important considerations are the need for the continuing availability of existing information products and the needed expansion of the information "product line" as scheduled.

GIS job descriptions

GIS job descriptions

The following is a list of sample GIS job descriptions. Most GIS job descriptions of the same titles are similar, however, they do vary depending upon the software used within an organization, the size of the system, the specific job responsibilities required by the position, and the type of agency or company seeking the position.

GIS manager

Provides on-site management and direction of services to develop, install, integrate, and maintain an agency-wide standard geographic information system (GIS) platform. This position will be responsible for developing, implementing, and maintaining special-purpose applications consistent with the agency's mission and business objectives. Requirements: must have proven experience in project-design and work-plan development; database system and application design; maintenance and administration of a large Oracle SDE database. Successful candidates must have a B.S./M.S. in geography, planning, or related field and three to five years professional experience implementing in-depth, complex, GIS solutions involving RDBMS and front-end application development. Applicant must also have knowledge of and ability to apply emerging information and GIS technologies (particularly Internet technologies); experience in project management; experience working with Oracle databases; and excellent interpersonal, organizational, and leadership skills. Experience with object-oriented methods and techniques is a plus.

Enterprise systems administrator

This position will provide user support, resolve UNIX- and NT-related problems, perform systems administrative functions on networked servers and configure new UNIX and NT workstations. Requirements include a B.S. degree in computer science or other related college degree with experience in computer systems area, or three or more years working systems administration functions in a client-server environment required. Must be familiar with multiple UNIX platforms and have high skills with both UNIX and NT commands and utilities. Good problem-solving skills and the ability to work in a team environment are mandatory.

GIS application programmer

The GIS programmer will design, code, and maintain in-house GIS software for custom applications. Position requires the ability to interpret user needs into useful applications. Candidate should have a B.S./B.A. degree or higher in computer science, geography, or related earth sciences. All candidates must have two years or more of programming experience with one of the following: VB, C++, or GIS vendor-specific programming languages. Previous experience with programming GIS applications and knowledge of GIS search engines is a plus.

GIS database analyst

Responsible for the creation of spatially enabled database models for an enterprise GIS. Typical tasks include setup, maintenance, and tuning of RDBMS and spatial data, as well as development of an enterprise-wide GIS database. Position is also responsible for building application frameworks based on Microsoft COM approaches. Position requires at least a B.S./B.A. in computer science or geography. Candidates should have a strong theoretical GIS and database design background and experience with Microsoft COM objects approach. Preference will be given to candidates with experience in modeling techniques used by the enterprise.

GIS analyst

Responsible for the development and delivery of GIS information products, data, and services. Responsibilities include database construction and maintenance using current enterprise GIS software; data collection and reformatting; assisting in designing and monitoring programs and procedures for users; programming system enhancements; customizing software; performing spatial analysis for special projects; and performing QA/QC activities. Requires a B.A. degree in geography, computer science and planning, engineering, or related field or an equivalent combination of education and experience. The candidate should have one to two years experience with GIS products and technologies, especially those currently used by the enterprise. The individual should have experience with various spatial-analysis technologies and knowledge of current enterprise spatial server engines is a plus.

Cartographer/GIS technician

Responsibilities include, all aspects of topographic and map production using custom software applications, including compilation from various source materials, generation of grids and graticules, relief portrayal, creating map surround elements, digital cartographic editing, text placement, color separation, quality assurance, symbol creation, and cartographic software testing. A successful candidate will have strong verbal and written communication and have a B.A./B.S. or M.S. degree (depending upon position level) in cartography, geography, GIS, or related field; experience and/or coursework in GIS software and macrolanguages, Visual Basic or graphic drawing packages; familiarity with remote-sensing/satellite imagery interpretation. Candidate should provide a digital or hard copy cartographic portfolio for evaluation.

Benchmark testing

Benchmark testing

A benchmark test is a comparative evaluation of different systems in a controlled environment. The test is used to determine which system can handle the anticipated workload in the most cost-effective manner.

Benchmark testing is appropriate if you are planning to procure a large-scale system. A benchmark test is not a demonstration of what a system can do. The objectives are to find out whether each system under test can perform the required tasks and to determine each system's relative performance with respect to these tasks. A benchmark test should be designed to verify that the system can perform the functions necessary to create the information products required in a timely manner. It should evaluate whether each system can handle the data required, produce the information products needed, and carry out the core functionality. The focus of the tests should be functions that you will use most often.

The test is also used for assessing the price of the proposed system compared to its performance.

If the system you're going to buy will cost $10,000, no vendor is going to conduct a benchmark test for you. The system is too small to make benchmarking cost effective. It may cost a vendor about $40,000 to put on a benchmark test. That is why such tests are generally conducted only for large system acquisitions, and they are getting harder to recommend to vendors as system prices fall.

During the planning process you have considered the following aspects of the anticipated workload:

- The information products created by the system in each of the first five years
- The relative importance of each information product to the daily work of the organization (the assigned priority) and the wait tolerance of each information product
- The system functions required to create each information product.
- The data required to create the information products
- The yearly volume of each data type required to create the information products

During the conceptual system design for technology you summarized and classified the functions required by the GIS in a functional utilization table and graph. This provides you with details of:
- the total set of system functions required
- the frequency of anticipated use of those functions
- the relative importance of those functions

With the function requirements already determined you now need to establish the throughput capacity of the system. Throughput capacity is the amount of work that the GIS must be able to carry out over a period of time. This will be a function of the combined capabilities of the system software and hardware.

Vendors participating in the benchmark testing should be provided with information on functional requirements and throughput capacity. They will then respond with details of a proposed system configuration—including the type and capacity of the equipment, storage devices, network capacities, and the number of input and product generation workstations—and the cost of system acquisition and maintenance.

They may also include descriptions of system capabilities, support services, company characteristics, and contractual commitments in any prebenchmark proposals.

A benchmark test is a compromise between exhaustive and inadequate testing. You should aim to keep the test as compact as possible while still allowing it to produce reliable results.

To minimize the testing effort placed on the vendors but still allow yourself to gather the information required, you can adopt the following guidelines:
- Send approximately 85 percent of the test data and all of the test questions to the vendors two months in advance of the test. This gives the vendors ample time to create the required databases and set up their systems to perform at optimum levels during the test.
- Make sure that each vendor receives the same data and test questions. This uniformity permits the testing of all vendors on an equal basis. It also permits you to create testing scenarios with known answers.

- Provide approximately 15 percent of the test data on the first day of the benchmark test to permit real-time observation of data entry and data updating.
- Choose a set of information products that are representative of your organization's needs including some that require rapid response, some that are mission critical, and some that are high-capacity business support applications.
- Carefully choose which information products to test to ensure that the proposed system has the capacity to create the other information products required by your organization.
- Test most thoroughly those frequently used functions that are required for the highest-priority information products. Less-used functions or lower-priority functions should be tested the least, but all functions should be tested at least once. Test functions that are not sufficiently tested during the creation of the selected information products in a separate section of the test specifically designed for that purpose.

There are other questions concerning the logistics of testing that you need to address. These are addressed in the following sections.

Where should the test be conducted?

It is usually preferable to hold the benchmark test at a site chosen by the vendor. They may choose their own company headquarters. However, if the new system will be embedded within an existing enterprise network, it is better for the client to provide an isolated network that emulates the multiple data transmission speeds encountered in their network with wide-area connections. If the client provides the testing site, the vendor should supply the proposed hardware at the client's site using the vendor-recommended protocols. If it is not possible to establish the network test at the client's site, the proposed network may be simulated at the vendor's site with provision made to closely monitor network traffic volumes.

When should the test be carried out?

The vendors should select the test dates. Once the testing dates are chosen by all the vendors, send the questions and materials to the vendors by courier at appropriate intervals to ensure that each vendor has the same amount of time between when they receive the materials and their scheduled testing date.

Who should manage the testing?

A benchmark test team should be used to manage the process and evaluate the results. This should be comprised of key client personnel, such as users, technical-support staff, consultants, and managers. It is important for the users to have a sense of ownership in the decision. It is also important to include technical support and management staff. They will be important allies during implementation and in providing ongoing support. Moreover, these different perspectives will help make the selection decision a better one.

Who should monitor the tests?

A subset of the management team, preferably not more than four persons, and preferably with two or more with experience with previous benchmark tests, should monitor the actual tests. This monitoring team should be the same for each benchmark test in the procurement.

Who pays for the testing?

The current practice is to share benchmark test costs between the client and the vendors. The client typically underwrites the cost of test preparation, materials, provision of an isolated network for testing, test monitoring, result analysis, and reporting. The vendors accept costs of database creation, installation of hardware on the isolated network (if required), and performing system tests.

Benchmark testing is conducted to ensure that the proposed system can perform the functions you need to generate your information products. If you do not test a proposed system adequately you could easily make a purchase with little value to your organization.

Your evaluation of the functional requirements should answer the following questions:

- Can the systems being tested perform the functions specified in the RFP?
- What is the relative performance of different systems on a function-by-function basis?
- What is the impact of the proposed system configuration and network utilization design on the ability of the system to generate the required information products in a timely manner?

The functions that you wish to evaluate should be identified in each section of any benchmark test guidelines provided to the vendors. Each information product has performance requirements (e.g., wait tolerance) specified.

A benchmark test team, together with anyone involved in preparation of data and answers for the benchmark test, should evaluate the functionality. The monitoring team is an important part of the benchmark test team.

The monitoring team should consist of a minimum of two and maximum of four persons, including: technical staff with previous benchmark-test experience or one or two experienced GIS consultants, plus one or two staff from your organization who have first-hand knowledge of the GIS analysis that is required.

To evaluate functionality, the benchmark test team need:

- all observations made by the monitoring team during the benchmark tests
- the indicators of system performance by function measured during the benchmark tests
- the results of the verification performed after the benchmark tests

Each function should be allocated a score using the scoring system from the table on the following page. This scoring system should be provided to vendors in the RFP. When the test is complete, the benchmark test team should discuss each function to arrive at a unanimous decision for the final scores.

Now you should assess how the production of information products will be affected by the functional capability. To do this, use your list of information products ranked by priority. Assign a 0-9 function score to each function invoked to make each product; determine the total and average score for each product; and determine the worst score for any of the functions required to make each product.

These results will be sufficient where there is a substantial difference in functional capabilities between the systems under evaluation in the benchmark test. It should be clear if one or more of the systems includes restricted functionality that prohibits the making of high-priority, frequently used products.

Where differences between systems are less pronounced, you should produce a graph like those shown on page 247 to help your decision making. Plot the highest function score obtained for each information product on the vertical axis against the priority ranking for each information product

Appendix B

Score	Function appraisal	Criteria
0	Outstanding	All of the qualities in appraisal totally integrated into an operational system—the best in the industry.
1	Excellent	Elegant, well-thought-out solution for individual functions—very fast and user friendly.
2	Very good	Fast and user friendly.
3	Good	Adequate and fast OR user friendly.
4	Satisfactory	Adequate.
5	Functional	Function can be performed. Needs minor improvement to speed or ease of use.
6	Functional with limita-tions	Function can be performed. Needs substantial improvements to speed or ease of use.
7	Partial only	New software development required for part for function.
8	Absent or not demon-strated	New software development required.
9	Absent and constrained	Impossible or very difficult to implement without major system modification.

(on the horizontal axis). Plot a separate curve for each system that you have tested.

If you obtain a graph with separate curves for each system, with no overlaps, then the preferred system on functional grounds is the one that is lowest on the graph.

If the curves on your graph overlap, you must make a judgment depending on the relative importance of high-priority versus low-priority products.

Throughput capacity is the amount of work that your system must be able to carry out over a period of time. To evaluate throughput capacity you should determine system performance for data input, information product generation, and network capacity utilization.

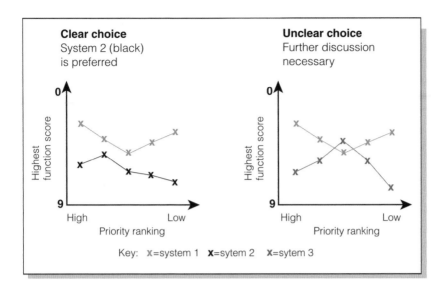

Data input

Calculate personnel time required for data input. When setting up your benchmark tests make sure you can extrapolate personnel time for each data set planned for input in each year of the five-year planning horizon. Determine data storage requirements: estimate these in terabytes, gigabytes, megabytes, or kilobytes for each data set as appropriate. When setting up your benchmark tests make sure you can extrapolate to include each data set by year for the five-year planning horizon.

Information product generation

Estimate the data volume by data function for each information product. Use the master input data list and the information product descriptions to generate this estimate. Use the measures of system performance by function and data volume that were collected during the benchmark test to calculate personnel time, CPU time, network capacity utilization, plotting time to generate each information product (if plotting is needed).

These results should then be compared with the results for the same variables generated during the benchmark test. After the benchmark test, you should be able to estimate by year the total personnel time required to input data, the amount of CPU utilization required to input data and create information products, storage requirements of the system, the network capacity utilization to create the required information products at the specified wait tolerances, and finally the plotter hours required for information product generation.

Compare the results of personnel time used for daily input and information product generation with personnel availability. Convert the results to operational costs by year using hourly rates so you can assess the overall use costs for each system.

The CPU utilization figures will require adjustment if the CPU type used by the vendor during the benchmark test is different than the CPU proposed for acquisition by your organization. Relative performance ratios provided by computer manufacturers should be used when necessary. In general, advise the vendor to use the proposed CPU for the benchmark test. Compare the results of the system resource use calculations with the system configuration proposed by the vendor. This will allow you to determine if the proposed configuration can handle the workload.

Network capacity utilization

When enterprise networks are needed for information product generation and tests have been conducted on an isolated controlled network, a quantitative evaluation of network capacity utilization is possible.

The benchmark test results allow you to calculate the impact on the network caused by use of the GIS. The benchmark test gives the percentage of network capacity utilization by information product for single- and multi-user conditions. These numbers can be extrapolated to estimate, by year, the network capacity utilization associated with the generation of the proposed information products. Estimates will reflect the hardware and protocol options and data search engine optimization design recommended by the vendor. You should compare these estimates with the client network administration numbers on network percent capacity availability in the same time period

The measures to be tested should be specified in the benchmark test guidelines and sent to vendors.

Network design
planning factors

Network design planning factors

Network design planning factors based on data per query analysis

Platform client	Data per query		Traffic per query	Bandwidth per user
	KBpq	Compressed Kbpq	Kbpq	Kbps
File server client	1,000	5,000	50,000	5,000
DC client	1,000	1,000	10,000	1,000
AS client	1,000	500	5,000	500
Terminal client	100	28	280	28
Browser client	50	50	500	50
Web GIS client	100	100	1,000	100

A system design performance factor is derived for six primary ArcGIS architectures, starting with average data requirements per query. Each of these analyses follows:

1. File server client:
This represents a standard ArcGIS desktop client accessing data from a file server data source. Data required to support display is 1,000 KB per query. Traffic between the client and server connection is 5,000 KB per query (complete file transfer required to access data extent required for display). Converting data to traffic yields 50,000 Kb traffic per query (8 Kb per KB with 2 Kb overhead per traffic packet). Analysis assumes one user query every 10 seconds, rendering average bandwidth utilization of 5,000 Kbps per user.

2. DC client:
This represents a standard ArcGIS desktop client accessing data from an ArcSDE data source, using ArcSDE direct connect to the DBMS data source. Data required to support display is 1,000 KB per query. Traffic between the client and server connection is 1,000 KB per query (no gain from ArcSDE compression). Converting data to traffic yields 10,000 Kb traffic per query (8 Kb per KB with 2 Kb overhead per traffic packet). Analysis

assumes one user query every 10 seconds, rendering average bandwidth utilization of 1,000 Kbps per user.

3. AS client:

This represents a standard ArcGIS desktop client accessing data from an ArcSDE data source, configured with ArcSDE installed on the server. Data required to support display is 1,000 KB per query. Data is compressed in ArcSDE®, reducing data transfer requirements to 500 KB per query. Converting data to traffic yields 5,000 Kb traffic per query (8 Kb per KB with 2 Kb overhead per traffic packet). Analysis assumes one user query every 10 seconds, rendering average bandwidth utilization of 500 Kbps per user.

Caution: The three architectures above are normally supported over an Ethernet LAN environment. Only one transmission can be supported over a shared LAN segment at a time; when two transmissions occur at the same time, a network collision results and the transmission fails. Each client delays a random amount of time before retransmitting the same data. Collisions due to too many users on a shared segment can result in rapid network saturation due to the retransmissions. For this reason, maximum traffic on shared Ethernet segments is typically reached around 20 to 30 percent of the total available bandwidth (due to frequent collision probability at higher utilization rates).

4. Terminal client:

This represents terminal client access to a standard ArcGIS desktop software executed on Windows Terminal Server. Data required to support display is 100 KB per query (display pixels only). Traffic between the client and server connection is 10 KB per query (average 10 percent compression with Citrix® ICA® protocol). Converting data to traffic yields 100 Kb traffic per query (8 Kb per KB with 2 Kb overhead per traffic packet). Analysis assumes one user query every 10 seconds, rendering average bandwidth utilization of 10 Kbps per user.

5. Browser client:

This represents browser client access to a standard ArcIMS image map service. Data required to support display is 50 KB per query (average size of image generated by ArcIMS service). Traffic between the client and server connection is 50 KB per query (no additional compression).

Converting data to traffic yields 500 Kb traffic per query (8 Kb per KB with 2 Kb overhead per traffic packet). Analysis assumes one user query every 10 seconds, rendering average bandwidth utilization of 50 Kbps per user. Note: If peak map request rates are given, 500 Kbpq would be a more accurate network design factor for design purposes.

6. Web GIS client:
This represents ArcGIS desktop client access to a standard ArcIMS image map service. Data required to support display is 100 KB per query (due to the larger client pixel display environment). Traffic between the client and server connection is 100 KB per query (no additional compression). Converting data to traffic yields 1,000 Kb traffic per query (8 Kb per KB with 2 Kb overhead per traffic packet). Analysis assumes one user query every 10 seconds, rendering average bandwidth utilization of 100 Kbps per user. Note: If peak map request rates are given, 1,000 Kbpq would be a more accurate network design factor for design purposes.

Caution: The three architectures above are normally supported over WAN environments. Only one transmission can be supported over a shared WAN segment at a time; when two transmissions occur at the same time, transmissions are cached at the router and fed sequentially onto the WAN link. Transmission delays will occur as a result of the cache time (time waiting to get on the WAN). For this reason, optimum performance on shared WAN segments is typically reached around 30 to 50 percent of the total available bandwidth (due to probability of delays at higher utilization rates).

Acknowledgments:
The author would like to express appreciation to Dave Peters, ESRI, for permission to use these network planning factors and the network traffic transport time and performance per CPU tables provided in chapter 10. For more detailed information, please refer to the "System design strategies" white paper referenced in the further reading section.

This lexicon is a reference to the most important GIS functions. Many of the function entries include examples of their use. This is not a comprehensive list of GIS functions, but rather an accessible, convenient subset of the most commonly used functions that will help you prepare information product descriptions and functional specifications.

There are several stages in the GIS planning process when you will discuss GIS functions:

- During the technology seminar (chapter 5) you need to discuss GIS functions to ensure that the planning team shares a common vision of what a GIS is and to encourage the use of a common vocabulary to assist communication.
- During the development of information product descriptions (chapter 6) you need to identify the GIS functions necessary to create an information product.
- During conceptual system design (chapter 7) you need to assess function use to help evaluate system requirements.

The lexicon is designed to:

- broaden your knowledge of the range of GIS functions available. It is estimated that most GIS users access only about 10 percent of the functions available to them on a daily basis. The lexicon provides an overview of the full range of functions that should be available in a comprehensive system.

- allow you to write a description of your needs in a way that will be understood by any user or vendor. Users and vendors of GIS software tend to become well-versed in the terminology of their particular software. Different software, however, may have different names for the same function or concept; therefore, a "software-independent" description will help understanding where terminology differs.
- allow non-GIS users to appreciate the capabilities of a GIS. Many of those involved in GIS planning will be new users. Senior management, clients, and members of an organization's information technology department may not have been previously exposed to GIS software. The lexicon will help you help them understand the full capabilities of GIS.

The lexicon is divided into these sections:

- Data input
- Data storage, data maintenance, and data output
- Query
- Generating features
- Manipulating features
- Address location
- Measurement
- Calculation
- Spatial analysis
- Surface interpolation
- Visibility analysis
- Modeling
- Network analysis

Functions that are considered high-complexity are indicated by an asterisk (*).

Data input

Digitizing

The process of converting point and line data from source documents into a machine-readable format. Manual methods employing a digitizing table or tablet are widely used but, increasingly, these methods are being replaced by scanning, automatic line following methods, and transferring data files already in digital format. Many users, however, still use a manual digitizing technique for small amounts of data or when other methods are too expensive. Digitized data often needs editing or reformatting.

Scanning

The process of creating an electronic photocopy of a paper map or document. You can use a flatbed or drum scanner, depending on the size of your map or document and the resolution you require. Scanning results in data that is in raster format (see raster-to-vector and vector-to-raster conversion). The data layer produced will contain all of the detail on the input map or image—including features you may not want to collect. For this reason, post-processing of scanned data is common. Scanned raster images are useful as a background for vector data, as they provide good spatial context and assist interpretation.

Keyboard input

The process of manually typing alphanumeric data into machine-readable form. This is infrequently used because the dangers of human error are high. Also, inputting multiple entries and checking are labor-intensive. Occasionally, keyboard entry is used to label data taken from hard copy maps or, less frequently, to input coordinates for map corrections or very simple maps.

File input

The process of using tabular data or text from ASCII or word processor-generated files that have already been manually typed and checked to input data.

File transfer

File transfer allows you to input data that has been created with some other software or system. The data file being transferred may come from external commercial data providers or from other systems within your own organization (for example, GPS or CAD). Data can be transferred from disk, CD-ROM, the Internet, or downloaded directly from field-data collection devices. This data might need to be reformatted to make it compatible with your GIS.

Lexicon

Raster-to-vector and vector-to-raster conversion

Processes that change the format of data to allow additional analysis and manipulation. During vector-to-raster conversion, you should be able to convert both the graphical and topological characteristics of the data. You should also be able to select cell size, grid position, and grid orientation. During raster-to-vector conversion, creation of topology is necessary. Editing and smoothing of data may be necessary to create an effective vector representation.

Data editing and data display

These should be possible at any time during digitizing and may apply to points, lines, and labels. On-screen displays or paper plots of errors assist editing. A range of functions for displaying different portions or different features of a data set is essential. Functions that should be available for edit and display include:

- selecting one or more data sets for display and editing
- selecting a specific area within a data set for editing
- selecting features (points, lines, labels, etc.) to be displayed
- selecting a feature or set of features to be edited
- querying the attributes of selected features
- indicating line ends (nodes) on request
- rotating, scaling, and shifting one data set with respect to another
- deleting all user-selected features, codes, or data types in specified areas
- deleting a user-selected feature using attributes or the mouse
- adding a feature using the cursor, keyboard, or mouse
- moving all or part of a feature
- creating and modifying areas interactively
- changing label text or location

Create topology*

The ability to create digitized lines that are intelligently connected to form polygons or networks. Ideally, the process is automatic and includes simple error correction procedures to join line endpoints within specified tolerance distances or remove small overshoots. More serious errors should be highlighted on-screen for you to edit and correct using standard graphical editing functions.

Edge matching*
An editing procedure used to join lines and areas across map sheet boundaries to produce a single seamless digital database. The join created by edge matching needs to be topological as well as graphical. An area joined by edge matching should become a single area in the final database, and a line joined by edge matching should become a single line. Edge matching functions should be able to handle small gaps in data, slight discrepancies, overshoots, and missed and double lines. If these gaps and errors cannot be handled automatically, you should be alerted to their presence so that other methods of correction can be applied. Edge matching functions should allow you to set a tolerance limit for automatic editing.

Adding attributes
The process of adding descriptive alphanumeric data to a digital map or to an existing attribute table. Attributes are characteristics of geographic features (points, lines, and areas). They are typically stored in tabular format and linked to the features. When there are a significant number of attributes associated with features, or sometimes for database design reasons, attributes may be stored in separate databases within your GIS or within another database management system (DBMS).

Reformatting digital data
The process of making digitized data or data transferred from another system compatible with your system. Reformatting ensures accessibility or assists conversion to the system software format. Reformatted data should be topologically and graphically compatible with other data in your GIS. You may need to reformat data so that it complies with organizational standards. Reformatting may involve additional digitizing, adding extra labels, automatic editing, and the editing of attribute data.

Data storage, data maintenance, and data output

Create and manage database
Database creation and management form the process of organizing data sets using good cartographic data structures and data compaction techniques. Database creation and management permits easy access to

data when a GIS contains several data sets, particularly if some of these are large.

In most cases, data will be input into a system from map sheets, images, or documents that cover large contiguous areas of land. The data from these sheets must be edge matched into a combined database with a consistent data structure. This will allow analysis and query functions to be applied to part or all of the database.

Edit and display (on output)

Edit and display of output map products on-screen are necessary to create effective information products. This process requires a wide variety of functions for editing, layout, symbolization, and plotting. Editing on output includes all of the editing capabilities used on input.

Appropriate symbolization assists presentation of results and simplifies interpretation of data. To facilitate symbolization, your GIS should include a wide variety of symbols that can be used to display points, lines, and areas; the ability to locate and display text and other alphanumeric labels; and the ability to create your own symbols

Symbolize

Symbolizing is the process of selecting and using a variety of symbols to represent the features in your database on-screen and on printed output. To create high-quality output from a GIS, you should use a wide variety of symbols to represent the features stored in the database on-screen and on printed output. Functions for symbolizing should permit the use of standard cartographic symbols, filling areas with patterns of symbols or cross-hatching of different densities, the representation of point features at different sizes and orientations, and the use of multiple discipline-oriented sets of symbols (for example, geological, electrical, oil, gas, water, and weather).

Plot

Plotting is the process of creating hard copy output from your GIS. Printing and plotting functions in your GIS should allow the production of "quick look" plots on-screen and on paper, spooling or stacking of large print jobs, plotting onto paper or Mylar® in a range of sizes, and registration facilities to permit overprinting on an existing printed sheet.

Update

Updating is the process of adding new points, lines, and areas to an existing database to correct mistakes or add new information.

After the initial creation of a digital database, periodic updates may be necessary to reflect changes in the landscape or area of interest. New buildings and roads may be constructed and old buildings and roads may be demolished. Quarries or forests may change in their extent or ownership, requiring updating of both the spatial extent and attributes of the data. You may need to update data to correct errors in the data.

Many of the functions for editing and display will be useful for updating, as will functions for heads-up digitizing on-screen. The ability to undo work is important, and there should be transaction logs, backup, and access protection for files during update.

Browse

Browsing is used to identify and define an area or window of interest that can be used for other functions. During browsing, no modifications to the database should be possible, but you should be able to select areas by specifying a window or central point, and pan and zoom. After identifying an area of interest, you should be able to edit, measure, query, reclassify, or overlay data.

Suppress

Suppression is used to remove features from your working environment so that they are omitted from subsequent manipulation and analysis. Unlike querying, which is normally used to select features you're interested in, suppression is used to omit features you're not interested in. For example, you might have a data source that contains all the roads in your study area, but you want to work only with the major highways. You can suppress all features other than the major highways so that they are excluded from subsequent overlay, display, and plotting operations.

Create list (report)

Creation of lists and reports may be used to generate final or interim information products.

Information products output from a GIS may be in the form of tables, lists, and reports. Functions in your GIS should allow you to create user-specified lists of the results of any function generating alphanumeric output;

produce subtotals, summary totals, and totals from lists of numbers; perform arithmetic and algebraic calculations based on given formulas; perform simple statistical operations, such as percentages, means, and modes; create list titles and headings in a range of standard formats and easily prescribed custom formats; sort data; and create reports of system errors to permit corrections to be made easily.

Serve on Internet

Serving GIS or map data over the Internet usually involves displaying interactive maps that allow users to browse geographic data. In addition, many map servers allow users to view attributes, query the database, and create customized maps on demand.

Query

Spatial query

Spatial querying is the process of selecting a subset of a study area based on spatial characteristics. The subset can be used for reporting, further study, or analysis.

Spatial queries are usually implemented by selecting a specific feature or by drawing a graphic shape around a set of features. For example, an irregular study area boundary may be plotted on-screen and all features within this boundary selected, or an administrative area might be selected with a single mouse click and used as a subset of the area for further study.

Querying a database can become complex and involve questions of both spatial and attribute data. For example, "Which properties are on the east side of town?" might be followed by "Which properties have four bedrooms and are available for sale?"

Queries are one of the most commonly used GIS functions. A good system will offer several alternative methods for querying to meet the needs of a range of users.

Attribute query

Attribute querying is the process of identifying a subset of features for further study based on questions about their attributes.

Attribute queries are usually implemented using a dialog that helps build the question or by using a special query language, such as structured query language (SQL). The questions "Which roads have two lanes?" and "Which properties are zoned residential?" would both result in the selection of a subset of features for further study.

Generating features

Generate features

Generating features is the ability to create new features and add them to the database. Generating functions should allow features to be defined easily with no limit on the number of new features that can be added to the database or on the number of points in any position. Names or codes can be attached to new features.

The types of features that you should be able to generate with your GIS include points, lines, polygons, circles, grid cell nets, and latitude-longitude nets.

Generate buffer

Buffer generation is the ability to generate zones of specified width around point, line, or area features. Around point and area features, these zones are generally called buffers, while zones of interest around line features may be called buffers or corridors.

A user-specified buffer distance is used to generate these buffers and corridors, and the system should automatically resolve overlaps and inclusions in cases where features are complex or highly convoluted. Buffers may be necessary both outside and inside area features such as lakes. For point, line, and area features, buffers at different distances (multiple buffers) should be possible. Constant and variable-width buffers should also be possible, including buffers that intersect each other. Buffer width should be able to be set from the attributes of the features concerned, without operator intervention.

Generate viewshed*

Generating a viewshed involves manipulating a digital elevation model (DEM) to identify areas of the terrain which are visible from one or more viewpoints. The viewpoints may be any points along a line (such as a road) or in a user-defined polygon.

Generate perspective view*

Generating a perspective view is the ability to generate a three-dimensional block diagram showing the nature of the surface relative to three axes from a digital elevation model. Hidden line removal, hill relief shading, and the ability to plot symbols and cross-hatched areas on the surface plane are desirable to achieve a good quality output.

Scene generation is an advanced form of generating a perspective view. It includes the ability to generate three-dimensional objects (for example, buildings, trees) and add them to the view to provide realistic visualization and to dynamically view the scene (perform fly-bys over and under the scene) and dynamically label the features on the passing scene.

Generate elevation cross section*

Generating an elevation cross section is the ability to generate a graph showing a cross section through a digital elevation model along a user-defined line of any length or orientation. It is useful if the locations of features that cross the line of section (for example, roads) can be annotated.

Generate graph

Graph generation is the ability to create a graph of attribute data. Graphs are used to display two attributes: one measured along the x-axis and the other along the y-axis. You should be able to illustrate data distribution with symbols, bars, lines, or fitted trend lines. Graphs can be used to supplement maps or drawn in place of them.

Manipulating features

Classify attributes

Classification is the process of grouping features with similar values or attributes into classes. Many data sets contain a wide range of values. Classifying the data into a number of groups, or classes, for presentation or analysis helps illustration and interpretation. For example, population totals for 1-kilometer grid cells may range from zero to several hundred in a study area. Displaying all possible values on a single map, using a different color for each one, would result in a multicolored map that would be almost impossible for the user to interpret. For presentation purposes, up to eight classes are normally used, but for analysis more classes may be appropriate.

Dissolve and merge

Dissolving and merging allows the removal of boundaries between two adjacent areas that have the same attributes. A common attribute is assigned to the new larger area.

This function may be necessary after an edge matching or reclassification operation (although in some cases it may be important to retain the boundaries between areas—for example, administrative or political boundaries).

Using these functions, the boundaries between adjacent areas with the same attributes are dissolved to form larger areas. Tables containing attribute values from the joined areas are merged to give one value for the resulting larger area.

Line thin*

Line thinning is the ability to reduce data file sizes where appropriate, following input, by reducing line detail. This function reduces the number of points used to define a line or set of lines, in accordance with user-defined tolerances. Some of the points along the line are "weeded out" to reduce the total number used to represent features. It is important that the general trend and information content of lines be preserved during the process.

Line smooth*

Line smoothing is almost the opposite of line thinning. It involves adding detail to lines to represent a feature more effectively. Line smoothing functions use tolerances to smooth lines by adding extra points and reducing the length of individual line segments. A smoother appearance results. A number of different functions for line smoothing may be available in a GIS.

Generalize*

Generalization is the process of reducing the amount of detail when displaying features. Generalization techniques are used to permit effective scale changes and to aid the integration of data from different source scales. A large-scale map (for example, 1:50,000) redisplayed on-screen at a smaller scale (for example, 1:250,000) will appear cluttered and difficult to interpret without the aid of generalization techniques.

Clip

Clipping allows you to extract features in the database from a defined area. This function is also commonly referred to as cookie-cutting. You may be able to define the area on-screen with the mouse or use another feature in the database (such as an administrative area) to define the area. The result is a new data layer that contains only the features of interest within your study area. The original data layer is not changed.

Scale change

Scale change involves changing the size at which data is displayed. A scale change can be performed in the computer rather than at the plotting table. Zoom-in and zoom-out functions should be available, as well as the ability to specify the exact scale at which you want to redisplay data. Line thinning and weeding operations or line smoothing may be incorporated in scale reduction. Line dissolving and attribute merging functions usually need to be invoked prior to broad-range scale reduction. Particular attention should be paid to the legibility of the final product, including labels.

Changing the scale of a data set before integrating it with other data should be undertaken with caution, as data is best manipulated and analyzed at the scale of collection. As a rule of thumb, you should avoid changing a data set's scale to more than 2.5 times larger or smaller than the scale of the original source if the data is to be used in analysis.

Projection change

Projection change functions allow you to alter the map projection being used to display a data set. You may need to change the projection of a data set to allow integration with data from another source. For example, data digitized from a map using the Universal Transverse Mercator projection will need to be changed to be overlaid by a data layer that uses an Equal-area Cylindrical projection. Functions for changing data between a range of common projections or map datums should be available.

Transformation*

Transformation involves the systematic mathematical manipulation of data. It is the process of converting coordinates from one coordinate system to another through translation (shift), rotation, and scaling. A transformation function is applied uniformly to all coordinates—scaling, rotating, and shifting all features in the output. It is often used to convert data in digitizer units (most often inches) into the real-world units represented on the original map manuscript. Data from CAD drawing files may require transformation to convert from page units to real-world coordinates and permit integration with other data.

Rubber sheet stretch*

Rubber sheet stretching is used to adjust one data set in a nonuniform manner to match up with another data set. If you have two data layers in your GIS that you need to overlay, or if you have one map on a digitizing table that needs to be registered to another map of the same area already in the system, rubber sheet stretching may be necessary. Rubber sheet stretching allows maps to be fit together or compared. Common points or known locations are used as control points, and the rest of the data is "stretched" to fit one data layer on the other. Rubber sheet stretching is also frequently used to align maps with image data.

Conflate*

Conflation aligns the lines in one data set with those in another and then transfers the attributes of one data set to the other.

Conflation allows the contents of two or more data sets to be merged to overcome differences between them. It replaces two or more versions of the data set with a single version that reflects the weighted average of the input data sets. The alignment operation is commonly achieved by rubber sheet stretching.

One of the most common uses of conflation is in transferring addresses and other geocoded information from street network files (for example, TIGER/Line files from the U.S. Census Bureau) to files with more precise coordinates. Many files contain valuable census data but may be deficient in coordinate accuracy. Because attributes are valuable, conflation procedures were developed to transfer the attribute data to a more desirable set of coordinates.

Subdivide area*

Subdividing is the ability to split an area according to a set of rules. As a simple example, given the corner points of a rectangular area, it should be possible to subdivide the area into ten equal-sized rectangles. The boundary of the area may be irregular, however, and the rules applied may be complex. The rules will allow factors such as maximum lot size and road allowances to be taken into account during subdivision planning.

Sliver polygon removal

A sliver polygon is a small area feature that may be found along the borders of areas following the topological overlay of two or more data sets with common features (for example, lakes). Topological overlay results in small sliver polygons if the two input data layers contain similar boundaries from two different sources. Consider two data layers containing land parcels that will be used in a topological overlay. One data layer may have come from an external source—perhaps provided in digital format by a data supplier. The other data layer may have been digitized within the organization. After they are overlaid, it is likely that small errors in the location of parcel boundaries will appear as sliver polygons—small thin polygons along the boundaries.

Automatic functions to remove sliver polygons are available and are commonly incorporated within both topological overlay and editing functions. You should have control of the algorithms used for sliver removal. In particular, you should be able to control the assignment algorithm that will determine to which neighboring polygon a sliver is assigned or how it is corrected.

Address locations

Address match

Address matching is the ability to match addresses that identify the same place but which may be recorded in different ways. This function can be used to eliminate redundancy in a single list (for example, a list of retail store customers) or, more frequently, to match addresses on one list to those on one or more other lists. Address matching is often used as a precursor to address geocoding. The user can specify various levels of matching probability.

Address geocoded

Address geocoding is the ability to add point locations defined by street addresses (or other address information) to a map. Address geocoding requires the comparison of each address in one data set to address ranges in the map data set. When an address matches the address range of a street segment, an interpolation is performed to locate and assign coordinates to the address. For example, a text-based data file containing customer addresses can be matched to a street data set. The result would be a point data set showing where customers live. The resulting points must be topologically integrated with the database and usable as new features in the database. The new features should be usable by other system functions in combination with the rest of the database.

Measurement

Measure length

Length measurement refers to the ability to measure the length of a line. Measurements in a vector database may be calculated automatically and stored as part of the database. In this case, length can be retrieved from the database with simple query operations. In other cases, measurements may be calculated after you click on a source feature of interest. For example, you might select two locations from your on-screen map, then ask that the distance between them be calculated.

Measure perimeter

Perimeter measurement refers to the ability to measure the perimeter of an area. Measurements in a vector database may be calculated automatically and stored as part of the database. In this case, perimeter can be retrieved from the database with simple query operations. In other cases, measurements may be calculated after you click on a source feature of interest. For example, you might select a field from your on-screen map, then ask that the perimeter be calculated.

Measure area

Area measurement is the ability to measure the area of a polygon. Measurements in a vector database may be calculated automatically and stored as part of the database. In this case, area can be retrieved from the database with simple query operations. In other cases, measurements may be calculated after you click on a source feature of interest. For example, you might select a land parcel from your on-screen map, then ask that the area of the parcel be calculated.

It should also be possible to calculate the area of a user-defined polygon. User-defined polygons may subdivide existing area features in the database. In this case, only that part of the area feature within the user-defined polygon should be measured. Interior areas contained within polygons (for example, islands within lakes) must be able to be measured and subtracted from the overall area of the feature. It should be possible to implement three levels of "stacking" in area calculations without operator intervention. For example, you should be able to measure the area of land mass in a polygon that includes a lake that contains an island on which there is a pond.

Measure volume*
Volume measurement is the ability to measure the amount of three-dimensional space occupied by a feature. Volume measurements can be calculated when surface digital elevation models of features have been incorporated into the database (for example, you could measure the volume of a mountain, the volume of a lake, or the volume of an aquifer).

Calculation

Calculate centroid*
The centroid calculation function calculates the centroid of an area (or set of areas or grid cells) within a user-defined region. It generates a new point at the centroid and automatically allocates a sequential number to each centroid in the region. This can be a useful technique for labeling polygons created during digitizing and is often performed automatically on new areas created by dissolve and merge or overlay operations.

Calculate bearing
Bearing calculation is the ability to calculate the bearing (with respect to true north) between two or more points in a database. This is a geometric calculation based on the relationships between features. You should be able to perform this calculation independently or in combination with

other arithmetic, algebraic, or geometric calculations in macroprograms and iterative procedures.

Calculate vertical distance or height

Vertical calculation is the ability to calculate the vertical distance (height) between two points in a digital elevation model. The calculation of vertical distance between two points should be possible wherever the points are located in the region covered by the digital elevation model.

Calculate slope

Calculation of slope (change in surface value) is the abilitiy to calculate the slope along lines, or the average slope of an area.

Calculate aspect*

Aspect calculation is the ability to calculate the compass direction towards which a slope faces. This function requires a digital elevation model and a user-specified area. The average aspect of the region should be calculated, weighted by the amount of land in each aspect category.

Calculate angle and distance*

This function is the ability to generalize the shape of a linear feature into a set of angles and distances from a starting point. The user should be able to set angular increments and constrain the calculation to any known point along the linear feature.

Calculate location from a traverse*

This function is the ability to calculate the route and endpoint of a traverse, given a starting point and directions and distances of travel. It should be possible to enter the resulting route and endpoint (a point or grid cell) into the database.

Arithmetic calculation

This function is the ability to perform operations such as addition, subtraction, multiplication, and division. You should be able to perform arithmetic calculations independently, perform arithmetic calculations in combination with algebraic and geometric functions, incorporate arithmetic calculations into macroprograms, change variables and components of algorithms, and establish iterative procedures.

Algebraic calculation

This is the ability to perform operations based on logical expressions. You should be able to perform algebraic calculations independently, perform algebraic calculations in combination with arithmetic and geometric functions, incorporate algebraic calculations into macroprograms, change variables and components of algorithms, and establish iterative procedures.

Statistical calculation

Statistical functions perform simple statistical analyses and tests on the database.

Increasingly, statistical functions are common in GIS software programs; however, for more sophisticated analysis, data may have to be transferred to other statistical packages. Statistical functions should allow you to calculate mean, median, standard deviation, variance, percentiles, cross-tabulations, and regression.

Spatial analysis

Graphic overplot

Graphic overplotting is the ability to superimpose one map on another and display the result on-screen or as a plot to see the intersection of the data sets. When you use graphic overplot, the data sets are not integrated in the database and no new data sets are created. This function merely produces a visual impression of the interrelationships between two (or more) data sets.

Graphic overplotting is commonly used to combine thematic data layers to give a context for interpretation. For example, you might display the boundary of your study area, the roads within the areas, land-use polygons, and rivers and lakes before performing queries or other analysis. Graphic overplotting can also be used to display the results of analysis in a way that aids interpretation. A land-use data set might be overplotted on a landscape surface to provide a three-dimensional visualization of the changes in land use across a study area.

Topological overlay*

Topological overlay of one map on another will produce new data as a result of the combination of two input data layers. The attributes of the two input layers will be combined into a new set of attributes for the accompanying output layer.

Three types of topological overlay are frequently used:

Point in polygon

Point in polygon overlay allows you to superimpose a set of points on a set of polygons, determine which polygon (if any) contains each point, and add the results to the database as attributes of the points. If a point is contained within a polygon, the attributes of that polygon are added to the point.

Line on polygon

Line on polygon overlay allows you to superimpose a set of lines on a set of polygons. Lines are broken at intersections with polygon boundaries, and the attributes of the polygon that each segment of the line crosses are added to the attributes of that segment.

Polygon on polygon

Polygon on polygon overlay allows you to superimpose two polygon data sets. The result is a topologically integrated version of the two input data sets that can be used to create a new output map or for further analysis. Polygons in the output map will have attributes from both of the input maps.

Adjacency analysis*

Adjacency analysis is the ability to identify areas that are next to (adjacent to) each other, particularly those that share a common boundary.

Connectivity analysis*

Connectivity analysis is the ability to identify areas or points that are (or are not) connected to other areas or points by tracing routes along linear features.

Nearest neighbor search*

Nearest neighbor search is the ability to identify individual or sets of points, lines, or areas that are nearest to other points, lines, or areas specified by location or attributes.

Correlation analysis*

Correlation is the ability to compare maps that show the same area, but that represent conditions in different time periods. Correlation can be a very useful management tool. Quantifying and explaining the differences between two maps requires determining and comparing the differences between them. Correlation is one method for doing this. It may involve using overlay techniques and statistical functions.

Linear referencing*

Linear referencing is the ability to associate multiple sets of attributes with any portion of a linear feature. These attributes can be stored, displayed, queried, and analyzed without affecting the underlying linear data's coordinates. Linear referencing models linear features using routes and events.

A route represents a linear feature, such as a city street, highway, or river. Routes contain measures that describe distance along them. These measures provide an explicit location for data that describe parts of the route. The attributes associated with any occurrence along the linear features are known as events. Events are stored in a tabular database rather than with the data's geometry; therefore, they do not affect the underlying spatial data. These events are accessed as needed from the tabular database.

Linear referencing allows the computation of locations of events on linear features based on an event table for which distance measures are available.

Surface interpolation

Interpolate spot height*

Interpolation of spot height is the ability to predict the height of any point in an area from a digital elevation model. A new point is generated with height as an attribute.

Interpolate spot heights along a line*

This function is the ability to predict heights along lines using a digital elevation model. For example, if you have a digital elevation model and a hydrology network, interpolation can be used to generate points along streams at fixed increments of height (for example, 10 feet) above a given point on the stream. The same technique could be used with other networks, such as roads or pipelines.

Interpolate isoline (contour)*

This function is the ability to generate lines showing equal elevation from a set of regularly or irregularly spaced point values. If the values are height values from a digital elevation model, contours will be created. If the point values represent pressure readings, isolines are created.

Interpolate watershed boundaries*

Interpolation of watershed boundaries is the ability to generate areas of drainage using a digital elevation model and a hydrology network. There are many terms used to refer to the areas of drainage, including drainage basin, watershed, basin, catchment area, and contributing area.

Visibility analysis

Line of sight*

Line-of-sight functions compute the points, parts of lines, and sections of polygons that are visible along a line between a given target and a point of observation. Line-of-sight calculation requires a surface. Commonly, surface data comes from a digital elevation model (DEM). If you were physically located at one point in your data set (say on top of a mountain), a line-of-sight calculation will establish whether you would be able to see from that point to a target point (such as a lookout tower on another mountain peak some distance away).

Generate viewshed*

Generating a viewshed involves manipulating a digital elevation model (DEM) to identify areas of the terrain which are visible from one or more viewpoints.

Modeling

Arithmetic modeling*

Arithmetic modeling is used to add, subtract, multiply, or divide the values of one or more input data sets to calculate the values for a resulting data set.

Weighted modeling*
Weighted modeling allows you to assign weighting factors to individual data sets according to a set of rules and to overlay those data sets and perform reclassify, dissolve, and merge functions on the resulting concatenated data set. This may be done to identify regions with specific characteristics (for example, zones suitable for development). In this instance, proximity to market may be given a higher weight in the modeling process than slope or aspect characteristics of the land.

Network analysis

Shortest route*
Shortest route functions determine the shortest or minimum value path between two points or sets of points on a network. The minimum value may be expressed in terms of, for example, cost or time. When complex network analysis is not required, shortest route functions may be sufficient for many users. Shortest route can be used on any type of network data, including transportation, river, pipeline, or cable networks.

Network analysis*
Network analysis functions allow you to perform a range of operations on network data. Shortest route and connectivity functions are simple forms of network analysis. More complex analyses are often necessary on network data for electrical, gas, and communications applications. The analyses that may be required include simulation of flows in complex networks, load balancing in electrical distribution networks, traffic flow analysis, calculation of pressure loss in gas pipes, optimization of complex delivery routes with tight constraints of time and load.

Further reading

Books

DeMers, Michael. 2002. *Fundamentals of geographic information systems.* 2d ed. New York: John Wiley & Sons, Inc.

Eason, Kenneth. 1989. *Information technology and organizational change.* London: Taylor & Francis.

Foresman, Timothy, ed. 1998. *The history of geographic information systems.* New York: Prentice Hall.

Harmon, John E. and Steven J. Anderson. 2003. *The design and implementation of geographic information systems.* New York: John Wiley & Sons, Inc.

Longley, Paul A., Michael F. Goodchild, David J. Maguire, and David W. Rhind. 2002. *Geographic information systems and science.* New York: John Wiley & Sons, Inc.

Mitchell, Andy. 1999. *The ESRI guide to GIS analysis volume 1: geographic patterns and relationships.* Redlands, Calif.: ESRI Press.

Muehrcke, Phillip and Juliana Muehrcke. 1998. *Map use: reading, analysis, interpretation.* Madison, Wis.: JP Publications.

Ormsby, Tim, Eileen Napoleon, Robert Burke, Carolyn Groessl, and Laura Feaster. 2001. *Getting to know ArcGIS desktop.* Redlands, Calif.: ESRI Press.

O'Sullivan, David and David Unwin. 2002. *Geographic information analysis.* New York: John Wiley & Sons, Inc.

Sommers, Rebecca. 2001. *Quick guide to GIS implementation and management.* Park Ridge, Ill.: Urban and Regional Information Systems Association.

Tomlinson, R. F. and M. A. G. Toomey. 1999. *GIS and LIS in Canada.* In Mapping a northern land : the survey of Canada 1947-1994, edited by Gerald McGrath and Louis Sebert. McGill Queen's University Press.

Zeiler, Michael. 1999. *Modeling our world: the ESRI guide to geodatabase design.* Redlands, Calif.: ESRI Press.

Web sites

The following Web sites include extensive reading on topics of relevance to GIS managers:

The official GIS home of the U.S. Geological Survey.
www.usgs.gov/research/gis/title.html

The U.S. Census Bureau's FAQ section.
www.census.gov/geo/www/faq-index.html

University of Edinburgh's GIS information clearinghouse.
www.geo.ed.ac.uk/home/giswww.html

Home of the leading GIS software.
www.esri.com

Journal articles

Calkins, Hugh W. and Duane F. Marble. 1987. The transition to automated production cartography: design of the master cartographic database. *The American Cartographer* 14 (2):105-19.

Tomlinson, R. F. and Douglas A. Smith. 1991. Assessing GIS costs and benefits: methodological and implementation issues. *International Journal Geographical Information Systems* 6:3.247-56.

Wilcox, Darlene L. 1990. Concerning "The economic evaluation of implementing a GIS." *International Journal of Geographical Information Systems* (April-June).

Wilcox, Darlene L. 2000. Now what do we do? Using cost-benefit analysis for strategic planning. *GEOWorld* 13 (2):42-4.

White papers

The following papers can be found in PDF format on the companion Web site to this book located at www.esri.com/esripress/tgis

A descriptive study of the usability of geospatial metadata. A reference to a usability study on the FGDC metadata standard done in Florida.

Building GIS catalogs and implementing a metadata catalog portal. ArcNews™ article and white paper on metadata and its role in GIS.

Building robust topologies. A paper on how and why ESRI implemented its particular topology format.

Migrating from ArcInfo workstation. Ideas and concepts in ArcGIS for ArcInfo users.

System design strategies. An overview of system design philosophy from the ESRI perspective.

logical data models, 129-146

logical linkages, 58, 72

M

master input data list (MIDL), 83
    basic components of, 84-85
    collecting information for, 88
    determining items needed for, 85

migration strategy, 191

mission statements, 22

ModelBuilder, 56

N

network analysis, 122

network performance
    effect of system configuration on, 157
    general issues related to, 158

network transport delay, 168

O

object-oriented data model, 137-142
    advantages of, 142
    components of, 137-138
    disadvantages of, 142

object-relational data model, 142-146
    advantages of, 143
    components of, 143
    disadvantages of, 144

organization of GIS project, 216-219

P

pilot projects, 193

planning proposal document, 27
    examples of, 29, 31

preliminary design document, 177

R

rapid prototyping, 56
relational data model, 129-136
  advantages of, 136
  components of, 131-135
  disadvantages of, 136
request for proposal (RFP), 205
  requirements of, 206-207
risk analysis, 194-197

S

schematics, 53
scope of GIS projects, 5-6
security review, 208-209
single-purpose projects, 5
spatial data, 3-4
staffing issues, 211-216
  GIS staff, 211
  system administration staff, 212
  training of, 214-215
standards, 116
steering committee, 217-218
strategic purpose, 21-24
survey, 117
system integration issues, 204-205
system procurement, 223-225
system scope, 91-104

T

technology lifecycles, 175
technology seminar, 35-42
temporal data, 119-120
text documents (as part of IPD), 52
timing of project
  factors affecting, 97
topology, 118

W

wait tolerance, 169
WAN (wide area network), 153
Web transaction processing, 156
workflow, 45

Thinking About GIS: Geographic Information System Planning for Managers
Copyediting by Tiffany Wilkerson
Book design, production, and image editing by Savitri Brant
Cover design by Takeshi Kanemura
Printing coordination by Cliff Crabbe